A Zoo Man's Notebook

A ZOO MAN'S

NOTEBOOK

By *LEE S. CRANDALL*

in collaboration with

WILLIAM BRIDGES

Chicago and London

THE UNIVERSITY OF CHICAGO PRESS

The University of Chicago Press, Chicago 60637
The University of Chicago Press, Ltd., London

© 1966 by The University of Chicago. All rights
reserved. Published 1966. Second impression 1975
Printed in the United States of America

ISBN: 0-226-11762-6 (clothbound); 0-226-11763-4 (paperbound)
Library of Congress Catalog Card Number: 66-13863

FOREWORD

I look upon *A Zoo Man's Notebook* as a leisurely stroll through a garden where I have spent my life—a zoological garden. Leisurely, for the work is all done now and the path is smooth. For nearly fifty years I had the daily care and responsibility, first for birds and later for mammals in the New York Zoological Park; and for ten more years I piled up sheet after sheet of yellow notepaper in the writing of *The Management of Wild Mammals in Captivity*. That book was written for the professionals, the men and women who now have the cares and responsibilities I had for so long. Necessarily, it records all the minutiae of the zoo business I was able to pack into it— the animals' habits in nature as they affect treatment in captivity, food, breeding, gestation periods, longevity, size, exhibition techniques, and the like.

But animals and zoos are too lively a subject to be expressed completely in the details of management. How a particular animal came to the zoo, what it did in the zoo, what we learned about it as a species or an individual are all part of the story. *A Zoo Man's Notebook* is essentially an abridgment of *The Management of Wild Mammals in Cap-*

tivity, but with the addition of such comments and anecdotes as seemed necessary to round out the accounts of certain animals.

I have not attempted to deal here with all the 1,382 kinds of animals discussed in the original book, but rather I have selected those species most likely to be familiar, by name at least, to almost everyone. Where some piquant detail happened to be recorded about a truly exotic animal, however little known, I have picked it out for the sake of exhibiting the variety and challenge of keeping wild animals.

Quite frankly, this is not a scientific book or a textbook; it is a "reading book." The general reader is not likely to keep an echidna or a gorilla or an okapi as a pet, and the details of distribution, distinction of races, and suggestions for exhibition housing would certainly not interest him. On the other hand, he can be expected to appreciate the complexity of the zoo world as various animals exemplify it and to come away believing that, as posters in the New York Zoological Park used to express it, "A Zoo Doesn't Just Happen."

ACKNOWLEDGMENTS

I am grateful to the New York Zoological Society, which not only gave me an opportunity to write *The Management of Wild Mammals in Captivity* but also through its president, Dr. Fairfield Osborn, and the Board of Trustees, permitted this condensation.

The Zoological Society's photographic library furnished the illustrations, and I am indebted to Mrs. Dorothy Reville, photo librarian, and Sam Dunton, staff photographer, for searching them out.

Although it is based on *The Management of Wild Mammals in Captivity,* this adaptation required additional writing. This was done by William Bridges, the Zoological Society's curator of publications (now retired) and my associate at the zoological park for more than thirty years.

L. S. C.

CONTENTS

viii   **Contents**

THE OPENING OF A NOTEBOOK

There is a good deal to be said for being the son of a doctor in upstate New York in the horse-and-buggy days. Often my father's practice took him over back-country roads, and except on occasions of emergency, I was generally welcome to go along and to fill the back of the buggy with anything I collected while he was making his calls. This was a valuable privilege, for I had pretty well exhausted the wildlife potential of the more readily available area—in other words, I had captured all the owls, turtles, and snakes I could find in the immediate neighborhood, or at least made them so wary there was very little chance of adding them to my private zoo.

These wild pets were a hobby, of course; my *business* at the age of fifteen was the breeding and exhibition of Sebright and Cochin bantams and I still have a printed letterhead with a bold line "Crandall and Company" to prove that I was a businessman. My father paid for the letterheads and thus kept me solvent.

It was taken for granted in the family that I would be a doctor. My grandfather had been a doctor, my father was a doctor, and naturally I would be a doctor too. This became official in 1907 when my family moved to New York City from Utica and I was enrolled in Cornell Medical School.

I stuck it out for one whole year, but my heart was not in medical studies. What I really wanted was something to do with animals—wild animals. This feeling was very real and ever present, but I didn't know what I could do about it. I suppose I knew, vaguely, that there were people called "naturalists," but I had never seen one or talked with one. In my world there were only doctors and medical students like myself.

In those days you did not lightly confess to your father that you were not enthusiastic about the career he had assumed you would follow, but I had no such reticence with one of my classmates. I confessed to him that I would much rather be working with animals than with sick people, but that there didn't seem to be any way of doing it.

This was no problem to my resourceful classmate. "I can fix you up, Crandall," he said, and proceeded to write a note of introduction to his uncle, William White Niles, who, it appeared, was attorney for an organization known as the New York Zoological Society, which was then in the process of building the New York Zoological Park in the Bronx.

Presenting a note to a busy lawyer in Wall Street struck me as a roundabout way of working with animals, but I had no better idea of my own and so I duly presented the note and in exchange got a courteous welcome and another note, this one to the Director of the New York Zoological Park, William T. Hornaday. Before presenting it I did a little research on Dr. Hornaday. He was a very great man, indeed. He had traveled all over the world collecting animals for Ward's Natural Science Establishment, he had done taxidermy for the Smithsonian Institution, and he had written books about wild animals. It took all my courage to invade his office in the zoological park and I was not at all confident that such a great man would have time to see me. Since then I have come to realize that even the director of a zoological park does not lightly disregard a letter of introduction from the attorney of his institution.

In any event, Dr. Hornaday listened to my account of the wild and domestic pets I had kept at home in Utica, my disenchantment with medicine, and my desire to work with animals. He was a decisive man. After very few questions, he announced that as soon as medical school was out that spring, I could come to work in the zoological park as a student, rotating duties in the mammal, reptile, and bird departments. He did not say anything about payment, and so I assumed that this glorious opportunity was free—I would not have to pay for the privilege he offered.

Thus it was in June of 1908 that I became a student in the mammal department of the zoological park. My duties were simple: I was to do what I was told to do. This did not vary greatly from building to building: cleaning cages, washing floors and windows, breaking open bales of hay, and carrying buckets of food were the major items. As a matter of fact, those are about the only duties an inexperienced student keeper is capable of carrying out safely.

Among other things, I learned how to do an impossible thing—walking unharmed through a red deer paddock dominated by a huge rutting buck in full antler. I never did it alone, then or since, but I did it alongside Keeper John Quinn. "Carry the bucket and stay close" was his only instruction, and then he seized a pickhandle and entered the paddock, banging on every rock he passed on the way to the feeding trough and back. Whether it was his boldness, or the banging, or the sheer unlikelihood of it all, so that even the savage buck could not believe his eyes, I do not know. At any rate, John Quinn entered the paddock and emerged alive day after day, and I stored the experience in my memory as one thing *not* to do. I still consider a rutting buck to be as dangerous as any animal in the zoo.

The climax of that summer was working in the bird house, for I had always wanted to try my hand at keeping wild birds but had no idea how to go about it. The prospect was completely fascinating. I was to work directly under Samuel Stacey, the head keeper, and since birds—at least Sebright and Cochin bantams—were my own specialty, Mr. Stacey looked upon me favorably from the first. His grandfather had been body sergeant to the Duke of Wellington during the Waterloo campaign and Sam grew up in the duke's household, where his father was water bailiff. The duchess herself took Sam under her care and made him her bird boy and eventually found him a place in the Zoological Gardens of London, whence he came to the New York Zoological Park to form and train a staff of bird keepers. There was a world of practical knowledge to be learned from Sam, but it had to be learned on the quiet, as it were. Ask him a direct question about why he was preparing food in such a way, or why he used this food instead of that, or how he hand-reared Argus pheasants, and you were likely to get an evasive answer or no answer at all. In those days gamekeeping was an inbred profession, filled with "secrets" that were—as we know today—hardly secrets at all. But they were part of Sam's hard-learned lore and he was not going to tell them to just anybody, just for the asking. On the other hand, look and listen and you learned everything that Sam knew, for he was a compulsive talker and could by no means keep a secret, even if he had had one. What he did know, to an amazing degree, were the everyday practicalities of taking care of birds, and many were the lectures I got on the right way and the wrong way as we worked along the back of the cages in the bird house, changing and cleaning food dishes, adjusting perches, talking to the birds, noting a molt that was not going fast enough, hunting delicacies for birds that were peckish, removing those that were unhappy with their cagemates.

It was Sam's delight, when we reached a certain turning in the angled row of cages, to stop and scratch his back against a projecting brick. "God bless the Duke of Argyll!" was his invariable exclamation as he rubbed and scratched.

I had been working in the zoo most of the summer when one day, as I was cleaning cages in the south end of the zoo, the curator of birds stopped to speak to me. I had seen William Beebe at a distance, of course, but never to speak to; student keepers did not waste the time of curators.

"Mr. Stacey tells me you like it here," was the way the conversation started. I don't remember any more of the exact words, but I still remember my feeling of gratitude to Mr. Stacey for having created this opportunity for me to actually speak with Mr. Beebe.

There was some questioning about my plans for the future. Was I going back to medical school? Did I think I would like a career in the bird department at the zoological park?

I must have expressed uncertainty about medical school, stalling for time until I found out what alternative might be forthcoming. When it came, it was beyond belief.

Mr. Beebe explained that he expected to be away a good deal in the future and would need an assistant. I could have the job if I wanted it. The salary would be $30 a month.

Such happiness comes few times in a lifetime; the human system could not stand repeated shocks like that. In the fall of 1908 I became a *salaried* employee of the New York Zoological Park. I have never regretted it.

THE ECHIDNA, OR SPINY ANTEATER

A zoologist is accustomed to thinking in terms of zoological classification, by which animals are grouped according to their relationships. So he begins a Notebook such as this with two of the most primitive (or less advanced) mammals, the echidna, or spiny anteater, of Australia and New Guinea, and the platypus, or duckbill, of Australia and Tasmania. Both are true mammals and suckle their young. And yet they lay eggs.

The echidnas have a fairly long history in captivity, dating from 1845 when the first one reached the Zoological Gardens of London. It lived for only four days, a precedent closely followed, it must be said, by most of its successors. But, most remarkably, not by all, for one echidna has since taken a position near the top of the list of mammalian life spans. It has been said that "the only mammals beside man that are known to exceed fifty years are the Asiatic elephant, and in very rare cases, the horse." Yet one echidna lived in the Philadelphia Zoological Garden from 1903 to 1953, a definite longevity of 49 years, 5 months, 15 days.

Few captive animals can have been so pampered as this one. Its life was spent in a modestly small cage, earth-floored

The echidna is a powerful digger. Released on hard-packed earth, this one dug itself out of sight in nine minutes.

and with an earth-floored nest from which it seldom emerged except at night or at dusk on a winter's afternoon. Never penned out of the nest box, it was free to run back into hiding at a sudden sound or at the sight of a visitor.

I saw this echidna first in 1909, a feat accomplished by inserting my index finger through the cage bars and gently lifting the lid of the sleeping box. I repeated the act for the last time in 1951, the only noticeable change during the forty-two-year interval being the slightly increased resistance of the rusting lid hinges.

Because of the strong, barbless spines and their defensive use, the echidna is one of the most difficult animals to handle physically. When threatened, the echidna can press itself to the ground or into a corner so tenaciously that it is next to impossible to dislodge it. At such times, the lower quills are turned strongly downward, so that an unwary hand may be severely injured.

I well remember two echidnas I was taking to New York from Sydney by steamer. These animals were inclosed in a strong wooden box and I had a daily struggle getting them to move so the box could be cleaned. I discovered finally that the only successful method was to tip the box so the animals would slide to the opposite side. There is no reason for supposing, of course, that production of a vacuum is in any way involved in their clinging ability.

The echidna has in its digging ability a fairly efficient escape method. Placed on an ordinary earth surface—even a rather hard one—an echidna, using the front feet with some assistance from the snout, can work its way below the surface in a remarkably short time. One in the New York Zoological Park disappeared from sight in hard-packed earth in exactly nine minutes.

But the echidna lives primarily above ground. The single egg is carried in a pouch, and the succeeding young animal is similarly transported until it becomes too large and troublesome to carry about. It is then deposited in some convenient hiding place where the mother visits it periodically.

THE PLATYPUS, OR DUCKBILL

The first skin of a platypus arrived at the British Museum in 1798, but it was not until 1884, when an English zoologist obtained eggs of both the platypus and the echidna in Australia, that the little duckbill was finally acknowledged as a primitive, egg-laying mammal. The protracted controversy concerning its nature is not surprising. Who *would* believe that a mammal laid eggs—and especially so improbably concocted a mammal, with silky fur, a broad and ducklike bill, webbed feet, and rooster-like spurs on its hind legs?

The broad, leathery bill of the platypus accounts for its alternative common name "duckbill." Although it lays eggs, it is a true mammal and suckles its young.

Considering the immense interest of such an animal, it is not remarkable that efforts were made to keep the platypus in captivity. In 1910 Harry Burrell, the "platypus man," began the experiments that laid the foundation of our present knowledge and finally resulted in the successful transportations of living platypuses to America. Burrell constructed a portable contraption that he called a "platypusary," consisting of a small tank and an attached labyrinth of tunnels through which the animal had to pass to reach its nest. The important feature was a series of rubber gaskets placed in the tunnels. These served to squeeze the water out of the platypus's fur as it passed through. Without them the nest would become wet, and the animal, unable to dry its coat, would be in risk of pneumonia.

Burrell's platypusary was well thought out, for it reproduced to some extent the animal's natural habitat. Semi-aquatic, the platypus seeks its food in the water—small crayfish, or "yabbies," aquatic insects, and worms—and then retires to burrows in the banks of streams for shelter and nesting. The platypusary had what the platypus wanted—a tank in which to feed, a tunnel, gaskets to scrape the fur dry as the sides of a natural tunnel would do, and a nesting chamber.

Some years after Burrell perfected his contraption, the late Ellis S. Joseph, a well-known dealer in animals, joined forces with him. In the spring of 1922 Joseph embarked aboard the *USS West Henshaw* with five male platypuses in a bulky platypusary and a large supply of earthworms for use as food for the animals. He reached San Francisco some six weeks later with one platypus but no earthworms, an early intimation of the enormous food capacity of these little animals.

After a stop to renew the food supply, Joseph proceeded to New York by train and reached the New York Zoological Park on July 14. Here the difficulties of maintaining the food supply immediately became apparent, and William T. Hornaday, at that time director of the Zoological Park, complained in print that the cost was from $4.00 to $5.00 a day. He gave as one day's ration ½ pound of earthworms, 40 shrimps, and 40 grubs—an amount that in the light of more recent experience seems hardly adequate. "Really, it seems incredible that an animal so small could chamber a food supply so large," he wrote. "I know of nothing to equal it among other mammals."

Each afternoon during its sojourn here the platypus was exhibited daily for one hour, long lines of visitors filing slowly past the open tank of the platypusary. The highly nervous temperament of the animal caused it to swim, scramble, and climb incessantly. On occasion, the keeper in charge did not hesitate to pick up the platypus and hold it—as long as he could—for the closer inspection of special guests. All of this, at the time, seemed perfectly reasonable, and the survival of the animal for forty-seven days was considered to be an unexpectedly good result.

In the next twenty-five years there were further experiments in Australia with keeping platypuses in captivity. In 1944 the Australian naturalist David Fleay actually succeeded in breeding the animal. To his pair, named Jack and Jill, a youngster named Corrie was born—or hatched!—after the extraordinarily brief incubation period of six to ten days.

Encouraged by this triumph, the New York Zoological Society decided to attempt another platypus importation. After a long wait in Australia for the construction of traveling platypusaries, accumulation of the necessary supply of earthworms, and, most important of all, securing of official sanctions, Fleay and his wife embarked for Boston in the spring of 1947 with one male and two female platypuses. After an arduous journey of twenty-seven days, during which the food supply was reinforced by the addition of fresh worms at Pitcairn and at Panama, the ship reached its destination, and the platypuses were convoyed to New York by motor. Three days later they went on exhibition in the New York Zoological Park in a permanent platypusary we had constructed for them.

It had all seemed so simple back in 1922 when the delicate and temperamental nature of the platypus was not recognized. Now that we knew more we realized that exhibiting a platypus even for one hour a day was not a simple matter. We soon found that 60° F. was the critical low temperature for the water in the tank; they liked it even warmer and below this point became increasingly reluctant to enter the water. A small electrical water-heating unit had to be installed in the platypusary.

When first allowed to enter the tank, the platypuses gave the usual indications of discontent, rolling over and over in the water and trying frantically to climb the corners. David Fleay considered that the disturbance was caused by the lack of overhead cover. We then arranged a green canvas, lined with white, above the pool to reduce the light, but the white lining had to be replaced with green before the animals were satisfied.

Fleay had named our animals Cecil, Betty Hutton (soon shortened to Betty), and Penelope. Because of the danger to Betty and Penelope from an injection of venom from Cecil's spurs, the male alone and the two females together were exhibited for one hour on alternate afternoons.

At 4:00 P.M. the slides that closed the burrow openings

were opened so that the animals could leave or enter at will, and their food was placed in the water. From dusk to dawn the platypuses were almost continually active, filling their cheek pouches under water and floating on the surface to break up and swallow the prodigious quantities of food consumed.

Normal daily rations for one platypus consisted of one pound of earthworms (about two hundred adult "night crawlers" and another large species of earthworm), two dozen live crayfish, one or two leopard frogs, two eggs steamed in a double boiler, and perhaps a handful of cockroaches or mealworms. All these items added up to at least double the amount of food that had been considered adequate in 1922. On this regime our three platypuses throve during the summer of 1947 and, by turnstile count, were viewed by just over two hundred thousand of our visitors.

A smaller version of the permanent platypusary was built in the basement of one of the zoo buildings for winter quarters, and the animals settled down quickly. When exhibition in the outdoor platypusary was resumed in the spring of 1948, Betty was obviously out of condition, although her weight remained at 1.98 pounds, her apparent normal. Gradually she began to lose, and in September of her second year with us she died.

Cecil and Penelope continued in the alternation of summers out of doors and winters indoors until the spring of 1951, when it became apparent that we should give serious thought to breeding possibilities. Always mindful of Cecil's venom glands and the spurs on his hind legs, we had built a separating wall between the male and female quarters in the winter platypusary. But eventually telltale tracks on the dusty floor revealed that one or both had found a way out through an unsuspected crack. When one morning Cecil was found asleep in Penelope's nest and Penelope curled up in Cecil's, friendly relations seemed obvious. Our hopes of duplicating Fleay's breeding success rose.

No behavior indicating anything beyond casual friendship and tolerance occurred, however, until the spring of 1953.

On May 21, at the close of Penelope's exhibition hour, she refused to return to her burrow as she customarily did. Cecil was then liberated to his side of the divided pool, in preparation for feeding, when Keeper Blair noticed that Penelope was scratching desperately at a wooden corner of the pool. Blair then removed the partition and the two animals almost immediately went into the circular mating maneuver that David Fleay had described for his Jack and Jill, Cecil grasping Penelope's tail firmly with his bill. When his grip finally loosened there was a brief pursuit, ending with Penelope floating quietly on the surface while Cecil preened her fur with his bill. This cycle was repeated four times up to 10:30 P.M., but no actual copulation was seen. For the next several days Cecil continued courting and Penelope showed no fright reactions, but the circling maneuver was not seen to be repeated.

David Fleay had written a precise and detailed timetable of the behavior of his platypuses before the appearance of young Corrie, and throughout the summer of 1953 Penelope's actions closely paralleled it. She began excavating the earth mound we had provided for her (corresponding to the bank of a stream in her native Australia), and when she showed disturbance at Cecil's company he was removed. She dragged dried eucalyptus leaves (obtained in quantity from the zoo's neighbor, the New York Botanical Garden) into her burrow, and finally she retired into the burrow for days at a time without emerging. When she did emerge she fed voraciously. Just as a test, in the early fall we gave her all she would eat: 2½ pounds of earthworms, 2 frogs, and 2 cooked eggs, no crayfish being available. This prodigious meal was completely consumed.

By Fleay's timetable, interpreted in the light of Penelope's behavior, her egg should have hatched on July 16. Fleay's Corrie emerged from the nest seventeen weeks after hatching, but we could not wait that long. Sixteen weeks from July 16 brought us to November 5, and cold weather with heavy snow was forecast for the following day. Penelope and her baby *had* to be dug out and removed to winter quarters.

And so the mound was carefully and methodically exca-vated. At the end of two hours of digging, Penelope alone was unearthed, with no sign of young, no proper nest, or even remains of the many leaves that she had carried underground from the pool. Why she should have paralleled Jill's cycle without result remains unexplained. The platypus is known to hibernate in some parts of Australia; we can only guess that Penelope's retirement in July was for hibernation rather than incubation.

After creating such a peak of hope and excitement, Cecil and Penelope returned to serene normality. Penelope some-how escaped from the platypusary on August 1, 1957, after she had been with us for 10 years, 3 months, 7 days. Cecil was found dead in his nest tunnel a few weeks later, after 10 years, 4 months, 24 days with us. Their life spans do not con-stitute a record for longevity—Fleay's Jack, with at least 17 years, holds that—but certainly they improved on our 47-day experience with the first platypus in 1922.

THE KANGAROOS AND WALLABIES

Of all the marsupials, or pouched mammals, certainly those most useful to the zoological garden are the kangaroos and wallabies. Most of the species breed readily in captivity, and from the moment the young animal first peeps from its mother's pouch until it is fully weaned, its development is easily observed.

Name distinctions, incidentally, are purely arbitrary in this group, "kangaroo" applying to the larger species and "wal-laby" to the smaller. Some species are known as wallaroos, others as euros.

The smallest is the charming and gentle gray wallaby of New Guinea, barely exceeding 2 feet in total length, whereas the largest is probably the gray kangaroo whose overall length is recorded at 9 feet, 7 inches; the red kangaroo is a close runner-up. Male kangaroos are greatly superior to females in

*Largest of the kangaroos is probably the gray, with a
recorded over-all length of nine feet, seven inches.
Old males are usually truculent.*

size and have the added, rather curious characteristic of continuing to grow long after they have reached maturity. Consequently, some huge old males, both gray and red, fully
deserve the common appellation of "Old Man." Such an
animal, of course, is usually not only truculent but actually
dangerous, since the powerful hind legs can deliver a blow
of crushing impact.

   In the New York Zoological Park, as in most zoos, kangaroos are kept in individual compartments each measuring 10
by 10 feet with the front of the cage and the partitions 7 feet

high. Most cage tops are open, but no kangaroo has ever leaped over a partition. Broad jumps of 27 feet and vertical leaps over barriers 9 and 10½ feet are reported, but such feats are accomplished only under great pressure not likely to be experienced in a zoological garden. For every two cages there is an outside run approximately 75 feet long by 20 feet wide, inclosed with wire netting 7½ feet high.

Keeping kangaroos in this manner is complicated by one serious drawback—the extreme nervousness of most species. Ordinarily, kangaroos are calm enough at close quarters but, like most animals that rely on running for escape from danger, they are likely to panic if suddenly frightened. We have lost several animals that have run into fences at top speed when startled by claps of thunder. There is danger, too, if a vehicle or other noisemaking apparatus should come suddenly upon the scene.

Our kangaroos are fed clover hay, rolled oats, feeding pellets, a mixture of ground grains to which salt and various minerals have been added, bread, cut potatoes, carrots, apples and bananas, and any greens available.

Under conditions of control such as individual cages provide, breeding becomes a matter of calculation. At the chosen time or when a male and a female in adjoining cages give evidence of special interest in each other, the door between the cages is opened, and the animals are allowed to run together until the female is known to have been bred. Various authorities give the gestation periods of the usual zoo species as from thirty days to six weeks. Because of the short span before the embryonic young emerges to enter the pouch, it is advisable to remove the male from the cage as soon as the female has ceased to be receptive. If closely confined with the mother following the birth, his attentions will almost certainly result in forcing the helpless young animal out of the pouch, so that, unless promptly discovered, it will dangle from the nipple to which it is attached, until it perishes.

After many years of controversy and misunderstanding, the facts of kangaroo birth are now established beyond question —facts which presumably apply, in general, to all marsupials.

Briefly stated, the expectant mother reclines low on her spine, with her tail extending forward between her hind legs, the upper level of the abdomen being nearly horizontal. The inch-long embryo emerges from the cloaca, climbs hand over hand along a strip of hair moistened with saliva by the mother, and enters the pouch unaided. Once it has found a nipple, the latter enlarges within its mouth, so that the embryo cannot be removed without damage.

By close observation an interested keeper can detect the small blood spots that indicate a birth—if cage floors are as clean as they should be—and exact dates of birth can often be determined. For this reason, I can vouch for the facts concerning a hybrid wallaroo-by-euro, born in our kangaroo house, which was seen to project its head from the pouch for the first time five months and eleven days after birth. For the next several weeks this young animal was in and out of the pouch until it had grown so large it could no longer enter. Even then, it put in its head to draw the nourishment necessary to sustain it until it could adapt itself entirely to the use of solid food. This schedule is, in general, that of all of the larger species.

Although the usual rule is for a single birth, multiple births sometimes occur. Among 219 births of kangaroos and wallabies in the Zoological Gardens of London, there were eleven pairs of twins and one set of triplets.

The greatest longevity for any kangaroo known to me is 20 years, 1 month, 22 days. This distinction belongs to a grizzled-gray tree kangaroo in the National Zoological Park in Washington.

THE BATS

The great majority of the nearly two thousand known forms of bat are unsuitable, in our present state of knowledge, for the zoological garden. But there are exceptions, notably the fruit bats or "flying foxes" of the Old World tropics and, to a lesser degree, the vampires and some others of Central and South America.

Bats in general have such highly specialized food requirements that it might be thought impossible to devise a broadly suitable diet. This has, however, been done for numerous small North American forms. And a most imaginative mixture it is, including chopped hard-boiled eggs mixed with cream cheese, cream, or milk, and occasionally chopped raw meat, vegetables, bananas, yeast, dry malted milk, Ovaltine, or finely chopped nuts. Cut-up honey bees are also sometimes added. Fruit bats, despite their name, are not narrowly restricted in their diet, for the Indian fruit bat has been reported as readily taking bread and milk, biscuits and boiled rice in addition to fruits.

Probably because of the ease with which their food requirements, consisting principally of fleshy fruits, can be met, the tropical Old World fruit bats do well in captivity. The larger species, which may have a wingspread of nearly 5 feet, are particularly attractive as exhibits. Although they usually spend the day suspended from whatever supports may be provided, they are continually active in small ways—shifting their wings, dressing their fur, or quarreling with their neighbors. Their large, bright eyes return the viewer's gaze with an apparent awareness not ordinarily expected in a bat.

Until the regulation was relaxed in 1960 for approved zoological gardens, the importation of fruit bats into this country was prohibited by law, on the basis that escaped specimens might become a menace to fruit crops in the warmer sections. In Europe and elsewhere the regulations were more tolerant and some remarkable longevities have been established by bats of this group. The Zoological Gardens of London kept an Indian fruit bat for 17 years, 1 month, 26 days, and the Giza Zoological Gardens maintained the African collared fruit bat for an astonishing 19 years, 9 months, 25 days.

If fruit bats are impressive zoological garden exhibits because of their size, the comparatively tiny vampire bats hold their own for another reason. Although local beliefs have

Some of the tropical Old World fruit bats have a wingspread of nearly five feet. Only since 1960 may they be legally imported into the United States.

attributed bloodsucking habits to bats of many kinds in many lands, the only species actually known to feed on the blood of other creatures are the members of one small neotropical family.

The first experiments in the maintenance of vampires in captivity probably were those made at the Gorgas Memorial Laboratory in Panama. It was found that if the vampires were provided with fresh blood that had been defibrinated to prevent clotting, there was no difficulty in keeping them alive.

In 1933 the late Dr. Raymond L. Ditmars, at that time curator of mammals and reptiles in the New York Zoological

The vampire bat of the New World tropics feeds exclusively on blood.

Park, brought back from Panama a single female specimen of the Panamanian vampire. Some two months after arrival the bat gave birth to a single young. Although the mother and infant died within a few weeks, Ditmars was able not only to confirm the previous observation made by the Gorgas Memorial Laboratory that the bat laps the blood with its tongue rather than sucking it, but also to demonstrate the fact by motion pictures. Subsequently two investigators working on Ditmars' bats showed that the flow of blood following the bite of a vampire is maintained by the sharpness of the incision and by licking the wound, rather than by the presence of an anticoagulant, as previously supposed.

The last vampire bat exhibited by the New York Zoological Park was kept under a blue fluorescent light in a glass-fronted cage. During the day it hid in the crevices of a section of log, but at 4:00 P.M., when a Petri dish containing one ounce of defibrinated blood was introduced, the bat promptly came down to feed.

The action of the creature on the ground is most interesting, since its wings can be so folded that it walks swiftly in
quadrupedal fashion or leaps with surprising speed and agility.

A zoo cannot, of course, allow its vampires to feed on living animals, and so the principal difficulty in keeping these
bats is establishing a certain and continuous supply of fresh
blood. It should be caught as it flows, in a sterile vessel, and
then whipped with wooden applicator or agitated with rough-
ened glass beads. The fibrin will adhere to the applicators or
the beads, and then, kept under refrigeration, the blood will
remain usable for at least a week. In older days abattoirs
abounded in New York City and fresh blood was readily obtainable—it used to be a routine sight for a huge delivery
truck to stop at the back door of the reptile house while the
driver delivered a quart bottle of blood once a week—but
since a supply of blood can no longer be readily maintained,
we do not attempt to keep vampires now.

EXHIBITING THE PRIMATES

If the perennial question, "What is the most popular animal
in the zoo?" is broadened to "What are the most popular
animals?" it can be easily answered: "The primates." They
compose the zoological order that includes man himself,
and although not many of the lemurs, lorises, galagos, pottos,
tarsiers, New World monkeys, marmosets, Old World monkeys, baboons, and great apes bear much superficial resemblance to man, there is no doubt that the universal interest
they arouse is based on the obvious human comparisons they
evoke.

Monkeys have been kept in captivity from antiquity to
modern times, and through the years housing arrangements
have grown from primitive inclosures to the stately buildings
of the present day. Oddly enough, when the use of glass to
insulate animals from visitors was introduced near the turn
of the century, there was much objection on the basis that

the animals pined when deprived of human contact. In those days before air-conditioning, it was also felt that the glass prevented proper circulation of air, and for a time glass generally was abandoned in favor of bars and wire netting. The use of glass has returned in recent years. Air-conditioning insures control of ventilation, temperature, and humidity; the close attention of well-trained keepers insures sympathetic treatment; and the risk of transmitting infectious diseases from visitors to animals is eliminated.

Glass does bring with it a variety of problems—one being that glass greatly increases the work of keepers. Various solutions for this important housekeeping problem have been developed, however, such as the use of glass on which an electric current is carried by a conductive coating on the inner surface. The charge is very light, but since few monkeys care to touch it more than once, the glass is seldom soiled. When the young are born, the current of that individual cage is turned off until the infant is old enough to learn its lesson.

Experience has shown that shock-resistant glass in itself is quite capable of restraining large and heavy animals. I have seen an angry male gorilla, weighing in excess of 400 pounds, drive against a sheet of laminated Herculite ½ inch thick and measuring 3½ by 4 feet, striking it with his shoulder at full force, with no result beyond the frustration of his ambition.

Although the use of glass is important in the control of atmospheric conditions within the animals' quarters, in improved visibility, and in elimination of offensive odors, its major value is in the reduction of exposure to tuberculosis and other contagious diseases of man to which many of the primates were especially susceptible. This is particularly true indoors; when primates are permitted the use of outdoor inclosures and the public is kept at a reasonable distance, the risk of infection appears to be negligible. Incidentally, exposure to the open air was once considered necessary to maintain good health in primates, but the practice now seems more sentimental than essential. Various primates living here in excellent condition have not been so exposed for ten years or more.

Methods of exhibition inevitably change, and this is certainly true of the ever-popular "monkey island." In the earlier stage unfortunate results often followed when large groups of baboons or monkeys were liberated on such islands without regard to sex or age. This practice led inevitably to serious fighting, heavy losses from mass infections, and, of course, an unfavorable impression on the public. But zoo men, like everybody else, learn from experience, and it has now been shown that there are at least three methods by which such exhibits can be operated with complete success.

First, there is the baboon group. Exhibition of these animals can probably be seen at its best in the Parc Zöologique du Bois de Vincennes in Paris, where about fifty adult Guinea baboons are shown in the approximate ratio of one male to ten females. Young animals are numerous and range from jealously guarded babes-in-arms to those already weaned. The latter are constantly active, riding on the backs of adults of either sex, staging mock battles, or tumbling each other about in riotous play. Among the mature animals there is no noticeable quarreling beyond local squabbles. The secret of success seems to depend upon the proportion of males to females, and upon the removal of any male that becomes overaggressive. Another solution is found at the Detroit Zoological Park where approximately sixty female Guinea baboons are maintained with but a single male. This certainly reduces the risk of fighting, and fifteen to twenty young are born annually.

Another ideal method is the family group, such as the colony of pig-tailed macaques in Artis, the zoological gardens of Amsterdam. In this group, a large and fine male is surrounded by perhaps a dozen females and young.

Finally, there is the "renting" system, probably practiced only in America. Under it, troops of young rhesus or other macaques, imported in the spring for eventual laboratory use, are farmed out for exhibition by dealers with the agreement that allowances will be made when the survivors are returned in the autumn. For the smaller zoos, lacking suitable winter accommodations, this plan assures a lively summer exhibit.

Whatever the method of maintenance, monkey islands are probably here to stay.

THE TARSIER

"The almost mythical tarsier" is the way I referred to this tiny primate in 1947 when, for the first time, I was privileged to peer through the wire screen of a shipping crate to see a pair of enormous eyes peering back at me from a rounded face— a face that seemed hardly large enough to contain such a pair of eyes. It was one of those rare occasions when zoo history was being made.

Until that time, very little was known of the natural history of the tarsier and even less of its proper treatment in captivity. The reference books revealed that the tarsier is distributed in a number of forms in the islands of southeastern Asia, from Sumatra, Borneo, and Celebes to the southern Philippines, but its small size (an adult fits comfortably in a cupped hand) and its nocturnal habits, coupled with the dread its bizarre appearance has aroused in native peoples of the area, made even the collection of museum specimens difficult.

Then, on July 9, 1947, Charles Wharton arrived in New York by direct flight from Manila with *thirty* tarsiers he had collected on the island of Mindanao in the southern Philippines. The "mythical" quality, it seemed, should be applied not to the animal's existence but to its rarity; Mr. Wharton happened to be collecting in an area that was being cleared for manila-hemp culture, and as the second-growth forest was cut away the tarsiers were concentrated and readily collected.

Thanks to Mr. Wharton's observations, a great deal is now known about the natural history of this minute—and distant —relative of man, but at that time there was a great deal to learn about its care in captivity. While he was in the field, Wharton had found that captive tarsiers would eat grasshoppers or locusts, mealworms, small crabs, lizards, bird flesh,

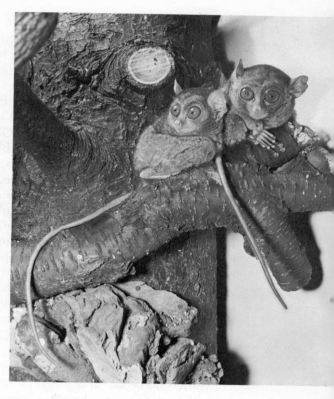

Tarsiers were believed to be extremely rare until a collector began working in an area in the Philippines where the forest was being cleared.

beef, liver, and mice. Eventually, he struck upon the idea of injecting egg yolk and vitamin compound into the abdomens of geckos immediately before these creatures, thus fortified, were fed to the tarsiers. This innovation was completely successful and served to land the animals in New York in excellent condition.

It is always a ticklish business modifying the feeding habits of animals. While in Mr. Wharton's care, the tarsiers had been accustomed to receiving injected geckos offered by forceps through the open doors of tiny cubicles. In our larger cages this method of presentation was difficult, and so a variety of food was placed in each cage, in the hope that the animals might feed by themselves. To our extreme gratification, during their first night with us, one tarsier ate three baby mice, one small frog, five large roaches, and six mealworms, and the other (of the pair we had retained) took one baby mouse, six roaches, and two mealworms. Obviously the feeding problem was not going to be insuperable.

We next undertook to introduce the animals to each other, putting them in a cage about three feet in each dimension. Released in strange quarters and uncertain in the rather bright light, which must have been hard on the eyes of such confirmed nocturnal creatures, they moved slowly over the floor, apparently unaware of each other, until their bodies happened to touch. They immediately fastened on to each other with both hands and feet in what seemed to be a death grip and went into a stiffened, convulsive state. After watching perhaps a minute of this frightening embrace, we separated the animals with some difficulty and placed them on the floor at opposite ends of the cage. As soon as they found themselves free of encumbrance, both leaped agilely to the branches that had been provided. There was no repetition of convulsive behavior, and the little creatures soon became friendly, often sitting together closely wrapped in each other's arms.

The Wharton collection of tarsiers was gradually dispersed to various zoological gardens or medical institutions and their widely varying longevity is indicative of how little we know about the ideal conditions of captivity. Our first pair died suddenly on the same night, having lived just under a year. A male and two females sent to the Zoological Gardens of London were kept in seclusion, except for viewing by privileged visitors, and settled down to a main diet of mealworms irradiated with ultraviolet rays to increase the vitamin D content. In one night a single animal ate ninety-nine mealworms. London set a longevity record of about two years and 10 months, the best up to that time.

But how to account for the extraordinary success of the

Zoological Garden of Philadelphia where one female lived for 11 years, 10 months, 23 days? This individual not only greatly exceeded the longevities previously recorded but also gave birth to two single young, both of which died on the same day through accidental injury. The Philadelphia curator's own guess is that the lack of mealworms in their diet may have had something to do with it. Contrary to the practice of other institutions, he never offered mealworms, which are a host to the horny-headed worm parasite.

Who knows? In any event, the diets offered to and accepted by most of the Wharton tarsiers are a matter of record, and if tarsiers ever again become plentiful in zoos, they will benefit by the collective experience of the past.

A HANDFUL OF MONKEYS

Both in a literal and in a figurative sense a monkey may be a handful. I am thinking particularly of a pygmy marmoset that came to the New York Zoological Park many years ago. Its shipping crate was a small, light, flimsily constructed wooden box scarcely larger than a cigar box, and as I balanced it on my hand I wondered whether it actually did contain a living creature. Slowly slipping back the lid, I saw a tiny, brown-and-black animal cowering in a corner, obviously uncertain of the fate that awaited it but in any event not likely to be ferocious. I inserted my hand and the marmoset stepped daintily onto my fingers, then into my cupped palm, and nestled there in all confidence while I lifted it out of the box for closer examination.

This minute inhabitant of the mighty forests of South America is the smallest of all the extensive group of New World primates to which the term monkey is popularly applied, and it is literally a handful—a scant handful. At the other end of the scale are the large and heavy baboons, particularly the great chacma baboon, largest of them all, from South Africa. It and indeed all the baboons are a handful of quite a different kind, from the point of view of an operational zoological park. They are strong, aggressive, and potentially dangerous animals with few of the amusing and winning traits by which so many monkeys attach themselves to their keepers and the public.

One of the New World monkeys that has long been a challenge to zoo-keepers is the howling monkey, largest of the American primates—an adult male red howler in British Guiana has been reported to weigh twenty pounds. The name, of course, originates from the monkey's booming voice, amplified by an enlargement of the vocal apparatus produced by development of the hyoid bone. The overpowering sound made by a troop of howlers, audible for great distances through the tropical forest, has been described as terrifying; but actually it always sounded to me like a distant storm, echoing the sound of wind and crashing trees. The best opinion about the purpose of this tremendous vocal accomplishment is that it has to do with "inter-group coordination," rather than with the frightening of enemies, as has sometimes been suggested.

Howlers are seldom seen in zoological parks, and this is apparently due to their temperamental inability to adapt to captivity conditions—commonly known as "moroseness"—as well as to the specialization of feeding habits. The food of howlers in nature consists largely of leaves, buds, and blossoms, supplemented by seed pods and certain nuts. Food of this kind is rarely obtainable in a zoological park.

Still, howlers can sometimes be kept with fair success. In the summer of 1950 the Zoological Society's collector, Charles Cordier, returned from Ecuador with an infant mantled howling monkey, male, bearing the fairly descriptive name of Ugly. This tiny creature was still being bottle-fed, but in a few days he began nibbling ripe bananas. Ugly grew rapidly and at an early age was provided with a small rough towel, changed daily. Like most young primates, he became deeply attached to this bit of cloth, using it not only as a cape but as a means of retirement, sometimes covering himself completely. After a period of several days during which Ugly

Smallest of all the New World primates
is the pygmy marmoset.

refused to leave his shelter even for food, the towel had to be removed. Prompt return to normal followed, and the towel was never given back to him.

Up to the age of two years Ugly welcomed visitors by creeping forward in a crawling position, growling softly. He then sought the proffered finger or knuckle, upon which he chewed frantically but gently. After the age of two, however, his temper became more uncertain, except toward his keeper. He had acquired the ability to roar, too, and did so on frequent occasions.

Between the ages of two and three Ugly shared his double cage, at intervals, with an adult female weeping woolly monkey, an animal of almost identical temperament. During these periods they ignored each other, never becoming friendly or actively quarreling. When the cage was approached, the first monkey to engage attention was allowed to enjoy it, while the other sulked at a distance.

As Ugly grew older, his temperament began to match his name, and at the age of three he became so irascible that he was considered unsafe even by his keeper. He lived in the zoo for nearly three years and ten months, certainly a respectable period for a "difficult" animal. Ugly's diet during this time was a complicated one containing cereal, bread, greens, vegetables, fruits, egg, canned dog food, and multiple vitamin concentrates—certainly a far cry from the normal food of howlers in the wild.

The aptly named spider monkeys that range from southern Mexico over most of tropical South America have comparatively slight bodies and disproportionately long and slender limbs so that they are strongly suggestive of furry spiders. Their delicate appearance should not be taken as indicating lack of strength, however, for an adult male is a powerful and sometimes aggressive animal. I can recall the escape of such a specimen that, when pressed against the ground in a catching net, raised the ring above his head without the least difficulty, in spite of my utmost effort to hold it down.

Spider monkeys seem reluctant to enter water but will certainly do so under some circumstances. We once placed half a dozen on a small island in one of our lakes, separated from shore by perhaps 35 feet of water about 18 inches deep. All went well for a few days. Then, one animal was found to have reached the mainland—a discovery made by a very much surprised, early arriving telephone operator who had the presence of mind to imprison the young animal in an empty trash can before calling for help. The monkey was replaced on the island and a careful watch was kept. This resulted in my being able to witness a mass escape as the entire troop walked through the water one after another, each monkey nearly upright and with one hand clutching its upraised tail near the tip.

No births of the diminutive pygmy marmoset have ever occurred in the Zoological Park, but several other species have bred and have furnished interesting examples of parental

*Deceptively fierce-looking, the little golden marmoset
is a gentle and attentive parent. Care of the young
is taken over by the father except at feeding times.*

care. With the golden, common, black-tailed, and cotton-
head marmosets, at least, the young are cared for by the
father, and they are surrendered to the mother only for the
frequent feeding periods. Usually, they can be seen curled
about the father's back or shoulders as he moves about.
Oddly enough, the pattern in the black-tailed marmoset
showed some variation from that of the others. Following the
birth of twins, the male seemed reluctant to take them over,

although he occasionally did so. When the young were three days old, the mother was seen to press them with her back against the wire of the cage top, to which they finally clung. During ensuing days they could frequently be seen in this position or buried under paper strips on the floor, varied with intermittent carrying by the father. In spite of this apparently unconventional behavior, both young were successfully reared.

Any monkeys may present difficulty to the zoo-keeper, but in general the Old World monkeys are far hardier in captivity than their relatives of the New World. They are also distinguished by a number of easily recognized characters, one of the most familiar of which is that although not all New World monkeys have prehensile or grasping tails, no Old World monkey has such a useful "fifth hand." Their tails, if present, are incapable of grasping.

Most commonly seen in zoos are the macaques, even though they are, with few exceptions, dull in color and to the initiated probably the least attractive of the monkeys. Their great advantage is that they are relatively easy to maintain and are always in good favor with the public, which finds their antics and grimaces in accord with the popular conception of monkey behavior.

One of the best known of the macaques is the rhesus monkey of India, which has been imported into this country by hundreds of thousands for medical research. Its hardiness is proverbial. In New York, where temperatures may reach 0° F. or even lower, we have kept the Japanese macaque out of doors all winter, with only an open shelter, and it proved quite indifferent to the cold.

The rhesus temperament is just as hardy as its physique. At one time we had a rhesus family, consisting of a very large male, a female, and a three-year-old daughter. Like most adult male macaques, this animal was powerful and aggressive, so that only a steel plate, almost hermetically sealed, prevented him from attacking the baboons and gibbons in adjoining cages.

Rhesus monkeys have been used to populate monkey

Japanese macaques are among the least attractive of the monkeys, although as babies they have an undeniable appeal.

islands for so many years that there is virtually an unlimited number of variations of the overall design of such islands, and little or no variation in the basic essentials. For macaques it is customary to provide a surrounding water space some 18 to 20 feet across, the thought being not how far the animal can jump but rather how far he cannot! The water is graduated in depth so that a shallow area is provided, with a depth of perhaps three feet at the outer perimeter. Here, a smooth concrete wall rises to a height of at least five to six feet above water level. This requirement is necessitated by the fact that certainly most, probably all, of the macaques swim like beavers and will escape from the water moat if it is at all possible. I recall an instance in which rhesus monkeys frequently escaped from an island where the outer wall rose

The brilliant red and blue face
makes the mandrill the most colorful
of all the baboons.

three feet above water level. Careful observation revealed
that escape was accomplished by one swimming animal
climbing upon the back of another, then leaping quickly to
the top of the wall from this flimsy support.

The Old World monkeys, as distinguished from the anthropoid or manlike apes, reach their peak of size and power in
the baboons. One of the standard natural history books of the
last century uses "hideous" twice in a single sentence to describe the mandrill, although nowadays we are more inclined to use the term "bizarre" in speaking of this West
African primate with its brilliant red and blue muzzle and

Although some baboons are often referred to as "bizarre" or "fearsome," the hamadryas baboon of southern Arabia is better described as handsome.

scarlet rump, whereas "handsome" seems appropriate for the hamadryas and "fearsome" for the chacma. But adjectives, after all, record only individual impressions of animals whose size, strength, and obvious defensive abilities are certain to arouse strong reactions in all who see them at close range.

Apart from a form of the hamadryas, which occurs in southern Arabia, the baboons are African in distribution. The eight currently recognized species cover most of that continent except the northern and northeastern area.

Baboons breed freely in captivity, with the exception of the drill and the mandrill, and it is one of the triumphs of modern zoo practice that mandrill births, at least, are becoming more common. The truculence of the male has often been a source of trouble, but we have found that in roomy quarters where he can get away from his mate and his offspring when he feels like it, the "old man" will usually make no trouble after the young are born, and will tolerate with gentleness the playful attentions of the infant as it begins to move about. In these circumstances, we have had a succession of mandrill babies in the New York Zoological Park. Although it is noticeable that a mother with a nursing baby or growing youngsters always keeps a wary eye on the "old man" and circles him cautiously when she detects storm warnings, there is plenty of room for the father to get out from under foot, and family life in the big compartment is, on the whole, quite amicable.

Obviously, there is still much to learn about the housing arrangements and feeding habits of many of the monkeys, but they are such a vital and interesting part of zoo exhibition that eventual success is inevitable. For instance, the exhibition of the curious proboscis monkey of Borneo remains a problem only partially solved as far as the zoological garden is concerned. The animal has frequently been kept as a pet in its own country, and specimens were available on the market many years ago; indeed, the first director of the New York Zoological Park, Dr. William T. Hornaday, as long ago as 1880 saw in the market in Singapore "a fine pair of Proboscis Monkeys at $100." It is difficult to believe that such animals —the male is spectacularly endowed with an extensible nose through which it makes its honking call—should not have reached Europe or America with some frequency, although the only early record I have been able to find is of an immature specimen in the London Zoo in 1902. The proboscis monkey's exclusive diet of leaves has always been the stumbling block. But in 1956 the San Diego Zoo received a young pair that thrived on boiled white potatoes and yams, peaches, bananas, apples, oranges, grapes, green corn, peanuts, bread, celery, lettuce, flower blossoms, and berries. The male lived a little more than four years. So the problem is not insoluble.

THE GIBBONS

Called by any of their names—the great apes, the anthropoid apes, the manlike apes—the members of this family generate the highest interest in primates as zoological-garden exhibits. It is not difficult to understand why. In the whole great order of primates to which man belongs, the gibbon, the orangutan, the chimpanzee, and the gorilla most closely resemble man in their anatomy and, in some respects, in their behavior. We may watch many of the lower primates—the monkeys, marmosets, baboons, and the like—with amusement and sympathy and even occasional quick recognition of some very human gesture or trait, but it is a superficial observer indeed who can gaze at a gorilla without some stirring of wonder and kinship.

Our subjective identification with the great apes is, admittedly, not very strong when it comes to the gibbons. They are certainly the least manlike externally, and the smallest of the family. Their home is in southeastern Asia and some nearby islands.

More completely adapted for life in the trees than are the other anthropoids, the gibbons move through trees with great facility, swinging by their elongated arms and getting tremendous momentum by releasing their holds at the most advantageous instant. Leaps of 30 and 40 feet from branch to branch are easily accomplished in this way. On rare visits to the ground they walk upright and even run, sometimes with surprisingly good speed. The same method is used in walking along lofty branches. Their ability to maintain balance in an upright position is highly developed, and I have often watched gibbons walking a tightly stretched wire, used as a tree support, sometimes holding food in their hands.

The loud, musical voices of gibbons—a penetrating whooop-whooop-whooooop—are well known to zoo visitors, or at any rate to those who make their visits in the mornings. It is reported that the calls of white-handed gibbons in the forests of Thailand were heard most frequently in the morning, less often in the afternoon, and seldom during the middle of the day. This timing appears to be maintained by captive animals, and on many a summer morning I have heard the ringing whoops of our family of gibbons as I entered the zoological park gate a quarter of a mile away. Almost invariably the calls declined as the morning wore on, and ceased well before noon.

No anthropoid appears able to swim naturally, and gibbons have a particular aversion to water and will not willingly cross or enter it. An extreme example is seen on Captain Jean Delacour's estate at Clères, France, which contains a number of small islands, each usually inhabited by a single gibbon. The islands are separated from the mainland by narrow strips of water hardly more than 10 feet across, yet they form impassable barriers. On one of my visits, I noticed that the branches of a tall tree growing on the mainland were so close to those of one on an island that they could be easily reached by a gibbon. Yet, Captain Delacour assured me that even at such a height above the water the island inhabitant had never ventured the crossing.

Some years ago we placed two white-handed gibbons on an island approximately 150 feet long by 75 feet wide in one of our lakes. Some 35 feet of water, 18 inches deep, separated the island from the mainland, but the gibbons have never attempted the crossing. Incentives at least to wade have not been lacking. There is a viewing area on the mainland and visitors frequently attempt to throw food to the gibbons. This usually lands in the water and if it floats close enough, the animals will gingerly retrieve it. Never, however, have they been seen to enter the water nor will they reach beyond safe distances.

Since leafy trees of various sizes grow on the island, the gibbons are completely at home and their remarkable brachiated leaps, musical voices, and upright walking habits combine to form an exhibit surprisingly attractive. During more than 20 years of island operation only two tragedies have occurred. In the absence of keepers a young white-handed gibbon just over a year old was seen by a visitor to miss his grip on an overhanging branch and fall into water not

White-handed gibbons on Gibbon Island in the New York Zoological Park, showing their method of drinking. (Photo by Lilo Hess.)

more than two feet deep, three or four feet from the island. While it floundered helplessly, its mother watched from the shore but made no effort to assist it. By the time keepers could arrive, the young gibbon had drowned. The second instance concerned another young animal, killed when a tree crashed during a summer storm.

A deep aversion to water, certainly not indifference, must have been involved in the instance of the mother that watched her offspring drown without attempting rescue, for gibbon mothers are notoriously solicitous of their babies. Infant gibbons, like many babies, soon learn to pick up bits of food while their mothers are eating, but the food is invariably taken away by the mothers before they can convey it to their mouths. One baby on which we kept records was nearly six weeks old before it left its mother's body for the first time and another month passed before she allowed it to crawl as much as a foot away. At a little more than five months, it was observed taking its first upright steps, holding tightly to one of its mother's hands. We recorded in the day book, rather anthropomorphically perhaps: "Today Junior took his first step."

THE ORANGUTAN

The orangutan, whose native name means "man of the woods," is found only in heavily forested lowland areas of Borneo and northwestern Sumatra. Just as uninstructed visitors to the zoo have no difficulty in distinguishing the gibbon from the other great apes on account of its small size—a big male weighs only about 20 pounds—so there is no mistaking the red-haired orangutan. The all-black chimpanzee and gorilla are completely different.

Progressive physical changes from birth to maturity are extreme in the orangutan, particularly in males, as anyone knows who has watched an orang grow from soft, clinging, affectionate, "cuddly" babyhood to gross maturity, when the tiny, deep-set eyes are framed by enormous cheek callosities and the baby throat has developed a huge, goiter-like pouch. Adult males are very much larger than females, but there are wide differences in the actual weights that have been recorded, all the way from approximately 165 pounds that one investigator reported as the average of thirteen adult males to the 450 pounds that our own overly fat orang named Andy attained at an estimated age of just over 13 years.

Andy, incidentally, was accurately weighed on a walk-on scales set in the floor of his cage in the Great Apes House of the Bronx Zoo, and the figure can be relied on. Guessing orang weights from external appearance can be deceptive, as we learned when the late Martin Johnson deposited with

"Andy," a big orangutan in the New York Zoological Park, found this gesture effective in begging food. His mate, "Sandra," like all orang females, was much smaller.

the zoo a very large male, with cheek callosities fully developed. Experienced members of the zoo staff estimated Truson's weight at somewhere between 200 and 300 pounds, but after his death four months later, his actual weight was shown to be only 127½ pounds. Flaring cheek callosities and long, concealing hair can make a lot of difference.

In contrast with all the other anthropoids, the orangutan is slow and deliberate in its movements. It is almost entirely a tree-dweller, drawing itself from limb to limb with unfailing caution by means of its long and immensely powerful arms. Just how powerful an adult orang is we learned on the memorable day when the Martin Johnson animal, Truson, came to the Bronx Zoo. Like all other newly received primates, he was quartered in the animal hospital for health examinations and tests before being caged in proximity to our other apes that were known to be healthy. It was summertime and Truson was liberated from his massive wooden shipping crate into an outdoor extension of one of the hospital's stoutest cages formed of cross-tied half-inch steel bars. Granted that the top of the cage had been exposed to rain for several years and a certain amount of rust might have been concealed by the frequent painting, we were quite unprepared for what happened when Truson, in the course of exploring his new home, climbed the bars, braced his hands and feet against the cross-ties, and humped his back. Boltheads and welded joints snapped apart like pretzels. Fortunately the zoo's machine shop was only a few yards away and hastily summoned machinists installed clamps and fresh bolts that discouraged any further ideas Truson might have had.

On the ground the orangutan usually progresses on all fours. It is perfectly capable of walking upright, however, waddling slowly on its short and bowed hind legs, usually with arms extended above its head.

Being tree-dwellers, young orangs are especially keen on swings and soon acquire great skill in operating them. Equally great skill is needed on the part of the zoo staff in constructing a swing that will unfailingly stand up under orang use. We have found even the heaviest rope to be useless, as it is too quickly unbraided by the animals. Chains may be dangerous, since a young orangutan easily becomes entangled and is then likely to hang helplessly until rescued. The only safe and successful swings we have been able to devise are made of rigid metal piping, hung at the top from stout eyes just large enough to permit free movement but with insufficient clearance for the insertion of an exploratory finger.

As so often happens when dealing with wild animals possessed of great strength, unlimited time, and persistence, we arrived at this rigid-pipe type of swing through trial and error. I well remember the frustrated exasperation of the zoo's head machinist who had to bind the ends of heavy manila cable with wire, then to affix chains to eye-bolts in the ceiling by means of bolts whose threads he battered until the nuts could not be turned (he thought), and finally to sheathe the chain in close-fitting lengths of iron pipe.

The orangutan does not seem to possess the inherent fear of water seen in the gibbons and may readily learn to wade in shallow water, but it appears to be helpless beyond wading depth. We learned this in a most frightening way in the days when we still believed a deep, water-filled moat was necessary to prevent the escape of orangs from their outdoor yards at the great apes house. To test the animal's reactions a young female orangutan named Sandra, then about four years old, was placed in the outdoor yard. A keeper was assigned to remain with her until the experiment had been completed, but an emergency called him away for perhaps five minutes. On his return, Sandra was missing, but a slight rippling of the surface of the moat explained the mystery. Leaping into three feet of water, the man quickly found the young orangutan fumbling around on the bottom, making no proper effort to extricate herself. First-aid treatment brought complete recovery, but until the water depth had been reduced beyond the danger point, the orangutans were never again allowed outside unattended.

THE CHIMPANZEE

Probably the best known of all the great apes is the chimpanzee—the hardy, adaptable, amenable chimpanzee. It has long been the darling of experimental psychologists and of the trainers of animal acts for zoos and circuses, and was even the "test pilot" of one of this country's early space rockets.

The chimpanzees are found almost all across equatorial Africa in forested areas from Gambia and the lower Congo

River in the west almost to the shores of Lake Victoria in the east. Living chimpanzees must have reached Europe during the eighteenth century, perhaps even earlier. The records are obscure and difficult to unravel, for sometimes it is impossible to be sure whether the animal under discussion was a chimpanzee or an orangutan. But in any event there is no doubt that the first chimpanzee owned by the Zoological Society of London arrived in 1836 and lived for a little over five months. London obtained another one in 1845, at a cost of £300, a very large sum for those days. In fact, it would be a large sum even at today's reduced value of the pound sterling, for the "going price" of chimpanzees nowadays averages only about $650.

It is a rule of thumb in the animal business that no matter how amenable and trainable a young chimpanzee is, by the time it reaches the age of seven it is no longer trustworthy. Exceptional and well-trained animals may perform safely and satisfactorily for a few years longer, but eventually those that are employed in animal acts have to be "retired" to non-performing exhibition. Such a chimpanzee was the huge male known as Jimmy who came to the Bronx Zoo from the St. Louis Zoological Park in 1945. His age was estimated at 12 years. No longer required to do strenuous acrobatics in the St. Louis Zoo's remarkable animal show, Jimmy settled down to a life of comparative leisure and indolence in our collection. His weight on arrival was 147 pounds, a rather high figure and well above the average. Later, as it became obvious that Jimmy was still growing although his figure remained trim, we made continued efforts to check his weight. Although there was no difficulty in moving him to a neighboring gorilla cage containing a built-in weighing device—when the rightful occupant was out of doors—nothing would induce Jimmy to trust himself on the platform. It was not until ten years after his arrival that a well-directed stream of water confined him to a small area and he took the risk. He was recorded at 190 pounds, the greatest weight then known to me for a chimpanzee of either sex. Just a year later Jimmy reached exactly 200 pounds.

At the estimated age of twenty-five years,
"Jimmy," a chimpanzee, attained
the record weight of two hundred pounds.

We have never, unfortunately, devised any practical way of testing his strength, but we can assume that it is very great. Comparative pulling strengths of chimpanzees and men have been tested elsewhere, and the ability of male chimpanzees was found to be comparable to that of large men. On the basis of pound for pound of body weight, however, the power of the chimpanzee was superior.

The agility of chimpanzees that makes them such excellent acrobatic performers is something the zoo man needs to keep in mind, even though he does not use his chimpanzees in animal acts. For example, our great apes' cages are lined with glazed tiles with the interstices narrow and smoothly fin-

*By taking advantage of minute depressions
in the interstices between smooth tiles, young
chimpanzees learned to climb the smooth, vertical
walls of their compartment.*

ished, yet, young and medium-sized chimpanzees are able to
climb these walls, in the corners, to the very top by clinging
to the almost imperceptible joints while facing either inward
or outward. This remarkable ability was first noticed by a
keeper in the great apes house, who encouraged it by smear-
ing honey high in the corners. The four young chimpanzees
occupying the cage at that time vied with each other in climbing up to the honey, and all became expert—a fact we re-

membered with misgiving the following spring when the time came for them to occupy a moated yard outdoors. This yard has two wall corners of brick and we expected that animals so skillful at corner-climbing would certainly undertake to negotiate them. We therefore installed electric barriers of the type used to restrain cattle, but curiously enough no chimpanzee ever attempted to climb those easily scalable corners. We can only suppose that there were so many other amusements and diversions available out of doors that corner-climbing was not attractive.

The possibility that an animal will escape is never very far from a zoo man's mind, and there is a considerable documentation on wall heights and moat widths that various zoos have found escape-proof—or not. For medium to large chimpanzees we have found that a shallow water-filled moat 14 feet wide ending at a smooth perpendicular wall 10 feet high is satisfactory. Just the same, we have never dared liberate the huge male, Jimmy, in that inclosure. All chimpanzees are naturally good acrobats, but Jimmy is a professional; his "star" turn in the St. Louis Zoo had been to leap high in the air from the back of one of a circling line of cantering ponies, turn a complete somersault, and alight on the back of the following pony. True, there would be no pony to use as a springboard, but we have a high regard for Jimmy's ability to think of a substitute if he got the chance.

THE GORILLA

As late as 1915 William T. Hornaday, director of the New York Zoological Park from 1896 to 1926, in writing a review of the history of the gorilla in captivity, took a dim view of the possibilities of the future. He said, "there is not the slightest reason to hope than an adult Gorilla, either male or female, ever will be seen living in a zoological park or garden. . . . It is unfortunate that the ape that, in some respects, stands nearest to man, never can be seen in adult state in zoological gardens; but we may as well accept that fact—because we cannot do otherwise."

These are not the negative words of a confirmed pessimist, but rather those of a rugged and determined realist whose unhappy experiences with captive gorillas had convinced him of the soundness of his position. Although he mentions the fact that a gorilla (the well-known Pussi) had lived in Breslau for just over seven years (1897–1904), he evidently considered this achievement to have little bearing on the problem. How could he have known that Pussi's record was actually only the beginning, that within less than 40 years a census would show fifty-six gorillas in captivity, and that in another ten years at least six had been kept for more than 20 years? Or that gorillas would actually be bred and reared in captivity?

It is not strange that early efforts to keep gorillas in captivity were carried out so persistently. Although scientifically described and named by Savage and Wyman as long ago as 1847, the gorilla still remains, in popular conception, a fearsome, almost mythical creature. This impression, largely fostered by the lectures and writings of Du Chaillu in the 1860's, perhaps will never be completely eradicated. No one can view a superb adult male gorilla, such as those that have been seen in various zoological gardens since Dr. Hornaday's declaration of frustration, without feelings of awe. It is only too easy to transmute him, in imagination, into the storied terror of the jungle, the abductor of native women. That he is ordinarily quiet, even inclined to lethargy unless disturbed or annoyed, is not easily accepted.

As it happens, the progress that has been made in keeping the gorilla in captivity has more or less coincided with field studies that have revealed some of the details of its life in nature. We now know, for example, that gorillas are accustomed to spending the night in nests grouped closely together on the ground or sometimes in low trees or shrubs. The nest-building habit may carry over into captivity. Two young animals in the collection of the Cleveland Zoological Park used to carry automobile tires each night to their sleeping shelves and line them with empty sacks to form a service-able nest. Oka, our adult lowland female, laboriously gathers dry leaves in autumn as they fall into her inclosure and forms them into a flimsy circle in which she sprawls. The two huge male mountain gorillas formerly living in the San Diego Zoological Garden used to build a nest of straw at night.

Reference to lowland and mountain gorillas calls attention to a distinction that zoological parks fortunate enough to have a mountain gorilla—by far the rarer of the two in captivity—are not likely to overlook. The lowland gorilla is widely distributed in the forested part of west central Africa; the mountain subspecies—generally the heavier of the two —is confined to the mountainous volcanic slope of the eastern Republic of the Congo (formerly the Belgian Congo) and it spills over into Uganda. Captive specimens of most animals are likely to be heavier than those same animals in the wild, but even the average for wild adult male mountain gorillas is impressive—458 pounds. Lowland males averaged 344 pounds, a significant difference. But these figures scarcely approach the best records for captive gorillas. The San Diego Zoo's male mountain gorilla Ngagi weighed 639 pounds just before his death, and his companion Mbongo is sometimes reported as weighing 660 pounds. The last definite weight recorded for this animal was 618 pounds, and in subsequent efforts to weigh him, the scales fluctuated from 645 to nearly 670 pounds because the platform of the scales was not able to accommodate the huge bulk of his body. Even the smaller lowland gorillas can astonish us by their size. The Berlin Zoo's Bobby weighed 577 pounds.

Adult gorillas are perfectly capable of standing upright but seldom walk in that position except for short distances when hand supports are available. They prefer to walk on all fours, the weight being carried on the flat of the feet or sometimes on their outer edges and on the knuckles of the hands. I have frequently seen our large female, Oka, take three or four running steps bipedally before dropping her hands to the ground. She is one adult gorilla, incidentally, that seems to enjoy climbing and she is often to be seen high up in the massive tree trunks provided in the outdoor inclosures of our great apes house. In nature the gorilla seeks most of its food on or near the ground, although the younger animals are good climbers and in play may even launch themselves in flat or downward leaps for distances of a few feet.

Breast-thumping is a common practice in gorillas. It usually consists of rapid blows of the open hands struck against the chest or abdomen. The performance is generally of short duration, for the animal must stand nearly upright, a position it cannot maintain for long without support. This gesture has often been interpreted as one of rage or defiance, but actually it appears to be an expression of well-being and is usually followed by activity of an inoffensive nature. Breast-thumping is frequently practiced by very young animals and may be seen in adult females. It is a gesture performed without warning; I have watched two-year-old gorillas happily wrestling and tumbling with each other, and suddenly one of them stands up, beats a quick but surprisingly loud tattoo on his small chest, and then drops on all fours to scurry away and play with a different companion.

It would be too much to say that today we know all about the care and feeding of gorillas in captivity, but we have certainly come a long way since 1855 when Wombwell's traveling menagerie exhibited in England the first gorilla seen alive in Europe—exhibited it as a chimpanzee, incidentally! It was not identified as a gorilla until an anatomical examination was made after its death. The first living gorilla to reach the United States appears to have been an infant that arrived in Boston in 1897, where it died after five days. The New York Zoological Park imported the second specimen, in 1911, and it did little better. Because of long transportation delays, the animal was in weak condition when it was received, and it died twelve days later. The only food it took during its brief stay in the zoo was the hearts of two small banana trees, sacrificed by the New York Botanical Garden, and the inner linings of banana peels.

Today's swift air transport, contrasted with the weeks or even months that used to be required to ship a gorilla from

Africa to the New World, undoubtedly is a big factor in today's greater success. But there is more to it than that. Two points of paramount importance have emerged from a century of experience: (1) the animal must be taken at an early age, and (2) it should become attached at once to a person, male or female, who will give it the care and security it would have received from its mother. The further need to accustom it to a variety of suitable foods is then readily satisfied. This attachment of the young gorilla to one keeper is usually transferable to another sympathetic person without a time lapse sufficient to cause loss of condition. Once the stage of dependent infancy has been passed, the psychological needs of the young animal naturally change and, in males at least, the attachment to the trusted attendant seems to become more tenuous as age increases. Adult females seem to retain a stronger feeling of dependence. This was certainly true of Oka who, long after reaching adulthood, is friendly, gentle, and can even be handled to a certain degree. For many years one of the Bronx Zoo's special treats for visiting dignitaries was the sight of Oka's keeper—a slight man, scarcely a third of Oka's weight—casually walking into the big gorilla's inclosure and sitting down beside her while she surreptitiously explored his pockets for the grapes she had learned to expect.

It has sometimes been considered that an animal companion, usually a young chimpanzee, should be provided for an infant gorilla. It has always seemed to me, however, that such a companion reduces the strength of the essential human attachment.

A large gorilla in an open inclosure gives an impression not likely to be forgotten, and since the animal's leaping and climbing abilities are limited, it would seem to be relatively easy to confine it without bars. The first solution to such a problem in the New York Zoological Park was to create an outdoor yard, similar to that provided for orangutans, with minimum reaching distances of 10 feet perpendicularly and 14 feet horizontally, and with a fronting water moat originally 6 feet deep.

At the time the yard was designed, much thought went into

the decision to include such a depth of water. The literature contained little evidence about the swimming ability of the gorilla, and the scattered references were conflicting. The best conclusion seemed to be that the gorilla fears and avoids deep water and cannot swim "without tuition or practice." We did know that gorillas may become conditioned to shallow water. Youngsters in the San Diego Zoo played in water up to two feet deep, and Phil in the St. Louis Zoo learned to leap freely into water four feet deep that rose to his neck when he sat quietly on the bottom. Phil was never seen to make any effort to swim, however. Oka, as well as other gorillas of ours, wade in shallow water when they want to retrieve something that has fallen into it, but they appear to derive no particular pleasure from it.

On May 13, 1951, we learned what the books had not told us. On the afternoon of that day, a Sunday, while hundreds of visitors stood in front of the moated outdoor inclosure watching the play of Oka and her male companion Makoko, the latter lost his balance moving along the edge of the moat and tumbled into the water, then 6 feet deep. Two keepers saw the accident from the visitors' rail but could not enter the inclosure from that point. Within minutes other keepers entered the yard from the building, and one, himself barely able to swim, dived into the water and found one of Makoko's hands, and hauled out the 448-pound animal. It was judged that immersion had lasted from 5 to 10 minutes, and all attempts at resuscitation failed, even though a police emergency crew arrived promptly and used all its apparatus and skill.

This fatal result of a trivial accident brings up several points of particular interest. Makoko had had no opportunity to become conditioned to water, so that the element encountered in his plunge was entirely strange to him. He sank from sight almost immediately, apparently making only futile efforts to extricate himself, and was completely submerged when he was found by the keeper. Certainly he made no effective attempt to swim nor did he seize the steel cables set along the moat side, under water, against such an emergency. We can

Even when she was fully adult, "Oka," a lowland gorilla, allowed keeper M. Quinn to play with her. He had taken care of her when she was a baby.

definitely take it as a fact, I think, that the gorilla, like other anthropoids, does not swim naturally; but of course we cannot rule out the possibility that an animal like Phil of St. Louis, so willing to sit in water up to his chin, might eventually learn.

One of the latest things to be learned about gorillas is their reproduction, for until 1956 there had been no captivity birth. The normal explanation of this situation is that in all the years of experience with captive gorillas, no mature male and female, with the possible exception of our Makoko and Oka, had ever lived together. They, but for Makoko's untimely death, might have been the first to produce offspring in captivity. On the basis of what we know now, we can presume that they were sexually mature, since Makoko was 11 years, 8 months old at the time of his death and Oka about 6 months younger. Moreover, they were strongly attached to each other. Because we feared that their great weight and boistrous play might result in accidental injury, they were allowed to be together for only a few hours daily. When the separating doors were opened, they would rush together, embrace, and then engage in a period of wild chasing, thumping, and mock biting in which the initiative might be taken by either. This might continue for as much as an hour and be resumed after a short rest period.

If our pair was not to be the first to reproduce in captivity, the great increase in the numbers of gorillas in zoos, from the late 1940's on, much improved the possibility of breeding, for many potentially compatible pairs were included. This expectation was justified on December 22, 1956, by the birth of the first lowland gorilla in captivity, at the Zoological Gardens of Columbus, Ohio. Little Colo was born after a gestation period of 257–59 days and was promptly abandoned by its mother. Fortunately it was discovered on the floor of the cage and was revived, with some difficulty, within a few minutes of its birth. Its weight at birth was approximately 4 pounds, 2 ounces.

Since then gorillas have been born in the Zoological Gardens of Basel and in the National Zoological Park in Washington. And there will be more.

THE ANTEATERS

If bizarre appearance and curious habits are enough to make a good zoological exhibit, the anteaters certainly qualify. In the New York Zoological Park's photographic collection we have a photograph, made many years ago, that bears me out.

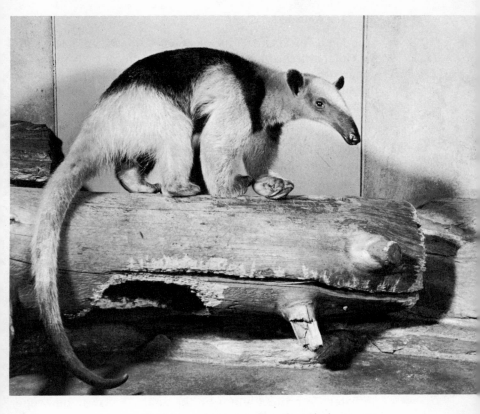

*The tamandua or prehensile-tailed anteater
of the American tropics is equally at
home on the ground or in trees.*

In that picture a straw-hatted man is staring at a giant anteater
with such a bemused expression that I always feel I know that
he is saying to himself: "Well, what do you know about that!"

The anteaters, the sloths, and the armadillos all belong to a
zoological order called the Edentata, or "toothless ones,"
which is not quite an exact designation, for both the sloths
and the armadillos have cheek or grinding teeth. But the anteaters do qualify, for they are entirely toothless.

The most spectacular is undoubtedly the giant anteater

found in tropical and subtropical regions from Venezuela and the Guianas to northern Argentina. Seen in captivity more frequently than either of its smaller relatives, the tamandua and the silky anteater, the giant's great size, strange appearance, and comparatively good viability make it sought after by zoos. Its habit of sleeping through much of the day is partially compensated by its curious practice of almost completely covering itself up with the long hair of its tail, so that, asleep, it resembles nothing but a low, rounded mound of grizzled yellow-gray-black hair. Zoo visitors have been known to point to a sleeping giant anteater and ask: "Is that something alive?" Among the leaves of a tropical forest floor, it must be almost invisible.

When the giant anteater walks, the long claws of the forefeet are turned inward, the animal supporting its weight on the outer surface of the toes. Ordinarily slow and deliberate in its movements, it is nevertheless capable of making fair speed at an awkward gallop. When the need for defense occurs, it assumes an upright position and is able to slash dangerously with its powerful claws. Sometimes the animal does not even have to assume an upright position, as a former veterinarian in the zoological park once assured me. He had been called to treat a small cut place on the mouth of a giant anteater. To protect him from the animal's claws, the keeper seized its tail and lifted it a few inches from the floor of the cage. The veterinarian then held the long muzzle, well away from the reach of the claws he thought, and dabbed the cut place with a stinging ointment. At the first touch, the anteater jerked, rolled sideways, and slashed forward with its left foot. The claws caught in the veterinarian's pocket and, as he said, "shelled out keys and small change like peas out of a pod."

The captivity diet of a giant anteater is a surprising one, considering that in nature it feeds principally on ants, termites, and perhaps stray insects of other groups that may be caught by the viscous coating of its probing tongue. The conventional diet for giant anteaters consists of milk, eggs, and chopped raw meat, which we modify by adding a small quantity of cooked cereal, such as oatmeal, and cod-liver oil or a few drops of multiple vitamin concentrate. We have never been successful in getting a giant anteater to take dried ant's "eggs" (pupae), dried locusts, or mealworms, for when these have been added to the mixture the sensitive tongue has always left them carefully at the bottom of the dish. The strangest variant on the conventional diet that I know was reported by the Zoological Gardens of Cincinnati, where a giant anteater that lived to a record age of more than 19 years devoured newborn mice with apparent relish!

Lacking teeth, which it does not need for feeding on soft-bodied termites and ants, the giant anteater draws food into its mouth by means of its long, protrusible, sticky tongue, which can whip through the liquid mixture of a captivity diet with astonishing speed. Because of the thickness of the mixture, it is sometimes hard to see just how much of the tongue is protruded during feeding, but it is quite noticeable during the animal's waking-up exercises. Then it is likely to swirl the broad, flat tail from its body, as if lifting a fan, and then rise and stretch its limbs, at the same time shooting out its tongue repeatedly. A keeper who had charge of our giant anteaters and had often seen this performance estimated that the tongue protruded at least 20 inches.

Baby giant anteaters have been born in captivity at least twice—in the Memphis and Columbus zoos—and both times the babies were killed by the male soon after birth. Babies have, of course, come into captivity with their mothers, and make most interesting exhibits, for they are perfect miniatures of their parent and habitually ride on the mother's back in such a fashion that their dark streaks blend with and continue the dark markings of the parent.

The tamandua or prehensile-tailed anteater is next in size between the giant and the diminutive silky anteater, but is considerably smaller than the giant. It is found from southern Mexico to Argentina and is intermediate in habits as well as size, for it is equally at home on the ground (like the giant) or in the trees (like the silky). In walking it turns the claws of the forefeet inward, as does the giant anteater, and adopts the same upright position in defense. Ordinarily it is a rather gentle and tractable animal, but its adjustability does not carry over into its diet. No one has yet come up with a formula that will keep it going for anything like the length of time that giant anteaters have thrived on the egg-meat-milk mixture. And yet, in nature, its feeding habits do not seem to differ greatly from those of its big relative; at least a pound of ants was found in the stomach of one specimen examined in the wild. There is still a great deal we do not know about the fine points of animal nutrition.

The smallest and least known of the anteaters is the tiny silky or two-toed, found from southern Mexico to northern Brazil. It is almost entirely a tree-dweller, although it can walk readily on the ground, with the two claws of the forefeet turned inward in typical anteater fashion. In the trees it moves with slothlike deliberation, and uses its grasping tail to insure safety. It appears to be more completely nocturnal than either of the larger species.

When alarmed, the little creature secures a firm hold on a branch with its tail and hind feet and draws itself upright, with the sharp claws of the forefeet held at either side of the face. In this position it presents a quaint and appealing appearance, but one that should not be misjudged, for it is in anything but a supplicating frame of mind! If approached too closely, it will bring its claw downward in a quick stroke, delivered with surprising force and power.

If the best longevity record so far for a silky anteater is just under a year and a half, there is good reason to believe that zoos can and will do much better. One specimen that came to us by air from Costa Rica settled down well and after the first day or two would assume the defensive attitude only if seriously disturbed. Contrary to expectation, it fed almost at once on a mixture of whole milk, minced raw meat, and coddled egg, thoroughly whipped up together. When feeding, it kept a firm grip on a low branch with tail and hind feet, leaning forward to place its front feet on the rim of the dish while the extensile tongue drew the food into its mouth. True, it died two months later, but examination showed that its abdominal cavity contained a large tumor. As anteaters go, the silky may eventually prove relatively easy to keep.

THE SLOTHS

Unless they know their animals, most visitors to a zoological park might have difficulty identifying a sloth without referring to the cage label. For the animal is usually seen as an amorphous mass of grayish-yellow hair hanging from a limb, no feature visible except the heavy claws by which it grasps the branch above the one on which it sits. Watch it long enough and you may see a drowsy shifting of position by an inch or two.

Naturally enough, the sloths of Central America and South America are firmly established in popular thought as symbols of laziness, yet they actually are extremely well adapted to their particular mode of life. Their food consists of leaves, buds, and twigs of tall forest trees, so that speed and agility are not required for security. Apparently defenseless, they are protected by dense coats of hair, highly resistant skin, and the ability to curl themselves into almost impenetrable balls. The green algae which grow on the hair, particularly during the wet season, serve as useful camouflage. Languid and hooking blows with the long claws of the front feet may inflict painful injuries, and the two-toed sloth is even capable of a serious bite, since the front tooth in each jaw is advanced and separated from the others.

Slow and deliberate as they are in any circumstance, they are completely at home in the trees—as we discovered on one occasion when a sloth was turned loose in a large pear tree growing in the New York Zoological Park. It was proposed to take a motion picture of the animal as it slowly crept along a horizontal branch within reaching distance from the ground. The sloth, however, had other ideas. It crept along the horizontal branch to the end and then without slackening reached up and grasped the branch above—and the next—and the next. Before the startled photographer could summon aid, the sloth was high out of reach in the top of the tree and the zoo's tree-pruning crew and its ladders had to be summoned to retrieve it. It was one of the few times that anyone in the zoological park had seen a sloth in action or moving more than a foot or two in its cage. No round-the-clock

The sloth spends its life hanging upside down in the tropical trees on which it feeds. Green algae grows on its hair in the wet season.

watch was ever kept, but it is likely that our sloths followed the same pattern as one kept in semiliberty in British Guiana. It averaged 18½ hours out of 24 curled up in sleep and moved about seeking food the rest of the time.

So completely are the sloths adapted to an upside-down life in the tree tops that on the ground they are almost helpless. They are able to progress on all fours, however, sprawling awkwardly and drawing themselves slowly forward, largely by the use of the forelimbs. Strangely enough, they swim with some facility.

The sloths fall into two natural groups, those with three toes on each foot and those with only two toes on the front feet and three on the hind feet. Three-toed sloths seemingly have such a strong preference for the large, thick leaves of the Cecropia tree that it is almost impossible to keep them in captivity for more than a few months. On the other hand, their relatives, the two-toed sloths, are easily maintained on greens, fruit, vegetables, and bread—the bread being moistened in water. This seems to provide sufficient moisture, for they seldom, if ever, drink water directly.

Sloths are certainly not exciting zoological exhibits, but some of them are persistent ones. A South American two-toed sloth lived in the Philadelphia Zoological Garden for 23 years, 2 months, 4 days.

THE ARMADILLOS

In popular articles on the defenses of animals (the bat and its "radar;" the skunk and "poison gas," and so on), the armadillos are invariably characterized as the "tanks" of the animal world. To make the comparison complete, however, we would have to endow tanks with the ability to dig themselves into the ground; even less capable of aggressive defensive action than the related anteaters and sloths, armadillos rely for protection on their plates of armor and on the surprising agility with which they reach the safety of their burrows or dig their way into soft soil. Some, although not all, are able to roll themselves into balls so tightly that they are practically impregnable. In fact, the three-banded armadillo closes its shell "like a steel trap" and apparently with comparable results, for it is seldom attacked by predators.

The armadillos of various species are found from southern Kansas and Florida to Patagonia, with the greatest concentration in northern South America. In size they range from the giant armadillo, sparsely distributed from Argentina to Venezuela, which may measure more than 5 feet in length, to the tiny fairy armadillo or pichiciego of Argentina, with a total length of only 5 inches.

Surprisingly, for an animal of confirmed terrestrial habits, armadillos are excellent climbers of wire netting—at least, they have no difficulty getting up, although getting down again is a different matter. Two nine-banded armadillos, which we kept in a large wire-inclosed cage, were found to be missing one morning and were later found sleeping soundly on a high ledge. They had climbed 8 feet up the netting and squeezed through a narrow gap at the top. Smooth metal plates 2 feet high, set at the bottom of the partitions, ended their wanderings. Our first giant armadillo died from injuries received when it fell from high on a wire cage front up which it had climbed. A barrier of widely spaced, perpendicular bars ended this danger for the next animal.

The giant armadillo is truly a giant among armadillos. One on which we kept a record of its weight came to us at 68 pounds and in something over two years had increased to 110 pounds on a single daily feeding of three pounds of raw chopped beef, five raw eggs, one quart of evaporated milk, one quart of tepid water, and one pint of tomato juice.

A giant armadillo is likely to impress a zoo visitor by its size, or what he can see of it, since the animal is likely to be half-buried in the earthen floor of its cage during the day. But at the coming of darkness, it puts on a fascinating show that was once described for us by the keeper of a large and very active specimen. The keeper spent the night in his building to observe what went on among his charges.

Anywhere between 9 and 10 o'clock each night the giant armadillo rouses from its deep sleep [he wrote]. Usually it begins the evening by a slow walk around the cage, sniffing and making what appears to be a general inspection. Quite frequently it walks in an upright position, using only the "pigeon-toed" hind feet. Although the animal does use the front claws for walking, nature obviously designed them to be most effective for digging.

The giant armadillo, ranging from Argentina to Venezuela, is sometimes more than five feet long. With this one, for comparison of size, is the more familiar six-banded armadillo.

After inspection is over, the armadillo begins its nightly dozen on the bars. It will lie perfectly flat on its belly, front claws extended. Then the left claw reaches across to the right to clasp a bar. When the curved claw has caught the bar, the animal will pull hard and do a complete side turnover, landing back on its belly. This performance is continued along the entire front of the cage, a distance of ten feet. When the steel side wall is reached, it reverses and starts back to the other side.

The floor of the cage is covered with eight inches of soil, so there is never a chance of it cutting its back during these rolling exercises. Although the shell of the giant armadillo has the appearance of being as hard as the shell of a tortoise, it is really quite easy to puncture, and bleeding results immediately.

After the performance with the bars, the animal starts on the shifting doors of its cage. It will insert its claws in the side of the steel door and pull until its body is brought down in contact with the steel side of the cage. In this position its great strength is brought into play. All the powerful muscles can be seen working with the precision of machinery, and the tail slaps against the steel wall with a force that reminds one of a pneumatic hammer. The building begins to sound like a boiler factory, yet never has the tail suffered from this constant pounding. After a few minutes, it walks to the bars at the front of the cage and, standing perfectly erect, proceeds to sharpen its front claws by drawing them down the bars. Then it starts digging. As fast as the dirt is loosened by the front claws, the hind feet push it away.

As daybreak begins, the almost incessant activity of the night begins to slow down. The armadillo gradually makes a complete circle. The head is slowly pulled under the shell, then the tail; in the meantime, the circling motion is continuing. The most remarkable feature of this roundabout route to a well deserved sleep is the way the animal slips its hind feet over its tail while making the circle. This it does without touching the tail, and so well is it timed that unless one keeps a careful watch, the performance will not be seen. When it is completely rolled up like a ball, all signs of activity cease.

THE PANGOLINS

The pangolins occupy a position in Africa and Asia comparable to that of the anteaters in the New World, although in a different zoological order. Toothless, their bodies, except the underparts, are covered with hard, sharp-edged scales, and their mouths are equipped with slender, extensile tongues. They are perfectly adapted for securing the ants and termites on which they feed. An alternative and apt name for them is the scaly anteater.

Our experience has been much the same as that of most other zoological gardens that have tried without notable success to keep pangolins, no matter how ingeniously they varied the diet. We had long wished to try, of course, without opportunity. I well remember eagerly awaiting the arrival of a ship from Calcutta bringing an animal shipment said to include a pangolin. This, however, proved to be an armadillo, which somehow had strayed far from home. Perhaps the confusion resulted because English-speaking people living in pangolin country referred to the animals as armadillos, just as sunbirds are often called hummingbirds, and "pangolin" and "armadillo" are sometimes used interchangeably.

Our first opportunity to gain firsthand knowledge of captive pangolins came in 1949 when our collector, Charles Cordier, arrived with a fabulous shipment of Belgian Congo

*By means of its long, sticky tongue the giant
pangolin, or scaly anteater, of Africa is
able to feed on ants and termites.*

mammals and birds. Included were one giant pangolin, one
long-tailed tree pangolin, and three white-bellied tree pangolins. The huge, terrestrial giant pangolin, measuring well over
4 feet in length, was practically moribund on arrival, refused
all food, and expired six days later. It was probably the first of
its species to be seen alive outside Africa.

Weakened as it was when it arrived, our giant pangolin had
no chance to demonstrate its strength, but a member of our
staff who had been in the Congo with Charles Cordier told
me about its vigor when it was newly caught. Notified by
telegram, Cordier drove 225 miles from Stanleyville to get
the animal. It had been penned in an up-ended oil drum on
the concrete porch of the village store. Since pangolins are
notoriously slow moving, Cordier ordered the drum lifted
when he was ready to dive on the animal and lift it off the
ground. But the pangolin was unexpectedly fast and hooked
its claws around a pole supporting the porch roof. In the end

it took six natives and Cordier to unhook it and stuff it into a cage for the trip back to Stanleyville.

Better results, although still certainly not satisfactory ones, were obtained with the slender, prehensile-tailed tree pangolins. All were quiet, slow moving, and inoffensive. If any area of the body scales was lightly touched, the razor-sharp scales closed instantly, and further urging caused the animals to curl into tight balls that could not be opened. Gentle handling could induce a demonstration of the powerful gripping ability of the tail tip, and the animal would hang suspended from one finger of its handler.

Shall we ever add the pangolins to the list of routine zoo animals? Perhaps. Even in the field, Cordier, for all his expertise, had many difficulties. Some refused food for several days and had to be released. Four specimens of the white-bellied and long-tailed pangolins showed general docility and willingness to feed on a "civilized" diet of reconstituted evaporated milk, precooked baby food, finely ground raw meat, coddled egg, and multiple vitamin concentrate, but 3 months and 3 days was the longest that any of them lived. At necropsy, all were found to be heavily parasitized. It is hard to believe that they were within three months of their death in the wild; therefore, captivity conditions were undoubtedly an important factor. Of all the pangolins, the giant —most spectacular of the lot—seems to have the best chance in the light of our present knowledge. I am thinking of the giant that went to the Fort Worth Zoo and Aquarium early in 1954, when it weighed 40 pounds and was 38 inches long. It was a young and rapidly growing female; a year and three months later it weighed 72 pounds and was 50 inches long. It lived for a little more than 4 years, to the amazement and envy of other zoos. What has been done can be done again, and some day I expect to see giant pangolins, and perhaps the lesser kinds, in every sizeable zoo.

THE HARES AND RABBITS

Anyone who has seen a furry creature with large ears and has exclaimed "Oh—here's a rabbit!" has a right to his confusion when he is informed that the animal is actually a hare. And confusion is natural enough, since the distinction between hare and rabbit is based largely on habits of reproduction. The hares, in general, do not dig or enter burrows, and the females give birth to fully furred young whose eyes are open. The European rabbit, on the other hand, lives in communities, sometimes very large, and digs and uses burrows in the earth; in them are born the nearly naked, sightless, and helpless young. The American cottontail and its relatives are intermediate in these respects, for although they do not dig burrows, they frequently make use of those abandoned by other animals. Their young, naked and sightless at birth, are born in shallow nests on the surface.

The European rabbit has been widely introduced in various parts of the world, usually with disastrous results. Its establishment in Australia, early in the nineteenth century, is a classic example of good intentions gone awry. The European fox, brought in to check the rabbits, has seriously decimated native fauna, including the lyre bird. Infection with a specifically lethal virus disease, myxomatosis, has now proved most effective in reducing the rabbit horde in Australia, but serious ecological damage is feared as a result of the intentional infection of rabbits in France with this rapidly spreading virus.

Wild hares and rabbits have never been kept commonly by zoological gardens, probably because the violence of their flight reactions is likely to lead to escape or injury and also because of their susceptibility to disease. In many ways this is too bad, for the great jack rabbits of the western plains are certainly part of the American tradition—as much so as the cottontail "B'rer Rabbit" of the Thornton Burgess stories familiar to a not-so-ancient generation—and zoological gardens might well feel an obligation to exhibit them. In the early days of the operation of the New York Zoological Park, jack rabbits were frequently received, but none survived

*The eastern cottontail rabbit is a familiar sight
in many zoos. Many of them roam wild in
the New York Zoological Park.*

for more than a few months. As I recall these animals, they
were wild-caught adults, with reflexes firmly established and
difficult to suppress. There seems to be no reason to suppose
that hand-reared jack rabbits, properly kept, should not do
very much better.

The eastern cottontail is more likely than any other to be
seen in American zoological gardens. Wild-caught adults do
not readily become reconciled to the conditions of captivity,
however, a fact that I once mentioned to a young newspaper
reporter without going on to explain that if cottontail babies
are taken while still nursing, they are easily reared on whole
cow's milk heated to body temperature and administered
with a medicine dropper or a small rubber nipple. The only
other requirement is that they be kept warm and sheltered.
The reporter wrote an article in which he declared uncondi-
tionally that it was impossible for the great Bronx Zoo to keep
a common cottontail rabbit in captivity. The volume of his

mail from farm boys who had reared baby rabbits to maturity was phenomenal.

Actually, such animals, if constantly handled and not frightened, will remain quite tame. Several are reared every year by the children's zoo in the New York Zoological Park, serve as attractive exhibits during the summer months, and are liberated in the grounds in autumn, or are carried through the winter and liberated in the following spring. Some captive cottontails have lived for 5 years, which is probably a good deal longer than guns and hawks would spare them in the wild.

THE RODENTS

The rodents are known in more forms than any other order of mammals; the accepted authority sets their number at the astonishing total of 6,400. Some are largely water-dwellers, some have taken to the trees or tunneled beneath the surface of the ground upon which the majority still live. Few land areas, from the Arctic region to the borders of Antarctica, have proved entirely inhospitable. Even Australia has its quota of rodents that, with the bats, are the only placental mammals indigenous to that continent.

Most rodents feed on plants in some form, but many will eat flesh if it is available, and a few are more directly carnivorous. Practically all are timid and depend upon flight for safety but usually will defend themselves fiercely if cornered. The larger species are able to inflict severe bites with their sharp incisors. Whereas some, such as most of the squirrels and the agoutis, are chiefly active by day, many others and particularly the smaller ones lead their active lives at night.

Understandably, night activity creates problems for zoological parks, whose visitors are present only during the daylight hours. Several ingenious systems have been devised to overcome this handicap—with varying success. The National Zoological Park in Washington induced some species to sleep at night by subjecting them to bright artificial light during this period, and exposed them in daytime to natural light filtered through blue-black or blue cellophane. The project was fairly successful, but the light was too dim for the animals to be seen well and still too bright to permit full daytime activity. The Chicago Zoological Park then experimented with blue light and carried activity-reversal one step forward. Probably the best results by the blue-light method have been obtained at the zoological gardens of Bristol and Chester in England by accomplishing the same reversal, but showing the animals by day under blue fluorescent tubes. This was reasonably successful for some species.

As these experiments progressed, the use of red light for the study of nocturnal mammals was increasing. Experimenters used an instrument known as a "sniperscope" that made use of a beam of infrared light to throw on a screen an image of the observed animal. Zoo visitors, unfortunately, want to see the animal itself, and not just an image of it. Then still another experimenter found that a flashlight, its lens shielded with a red acetate disk, allowed full nocturnal observation without disturbing the activities of the subject. This was the final key to the solution of the problem.

Reversal of day-night activity had never been attempted in the Bronx Zoo, our only efforts having been to protect the nocturnal species from glare by the use of blue fluorescent tubes. Then a series of experiments by our Curator of Mammals, Joseph A. Davis, Jr., soon convinced him that red light instead of blue gives the desired result. The rods that predominate in the retinas of the eyes of nocturnal animals are insensitive to color, so that red light is not perceived as such. Under it, animals conduct themselves as though they were in almost complete darkness.

Acting on this principle, we created a "red light room" in our small mammal house. Red fluorescent tubes were installed and even the cage walls were painted red. There is no light in the public space except for the red reflections from the cages and scattered red tubes overhead, but once visitors' eyes have become accustomed to the change, visibility is excellent. Under these conditions the animals are active, indifferent to the public, and plainly visible. So far no small mammal has been resistant to reversal, and success has been so striking that the idea is to be expanded in a new building, the World of Darkness, for the exhibition not only of small mammals but of nocturnal birds and reptiles. Such an exhibit could—indeed, it will—open up a huge and hitherto almost unknown segment of animal life to the public.

THE BEAVERS

There may well be 6,400 forms of rodents, but two of them have been and probably still are of greater economic importance than any others. These are the beavers—the Old World beaver and the American beaver. As a matter of fact, the American beaver is the only one that really counts, for the Old World species, which once extended across northern Europe and Asia, has for so long been persecuted for the value of its fur that it has largely disappeared from its former haunts. The American beaver once appeared doomed to a similar fate, but conservationists have been so successful in reestablishing colonies that the pelt harvest is still thriving.

Valuable as its fur is, general public interest is based on the animal's real and fancied abilities in tree-felling and dam-building. These are not easy to demonstrate in a zoo, however.

A beaver pond was one of the first projects undertaken in the construction of the New York Zoological Park at the end of the last century. A 4-foot-high fence, constructed of ⅜-inch iron rods, with an inside overhang at the top, was erected. This rested upon the coping of a concrete wall extending 4 feet into the ground and theoretically connecting with the underlying bedrock. This barrier enclosed an area of about half an acre, including a shallow natural pond covering about half the space. The newly installed beavers were soon busily engaged in felling the few small trees available so that finally only two or three larger ones remained, protected at their bases by guards of heavy wire netting. After the animals had stripped the edible bark from the branches, they used the branches in the construction of a lodge about 12 feet in diameter and 3 to 4 feet high. This lodge provided shelter for many successive beaver groups.

The encircling fence was designed to be escape-proof and actually was for many years. Eventually an exploring beaver found a spot where the concrete footing did not connect with the bedrock, and the first we knew of it was a report that "vandals" were cutting down some trees outside the fence. Before too much damage was done, however, we rounded up the four-footed vandals, sealed their escape route, and restored the colony inside the fence.

The possibility of escape apart, the exhibition of beavers involves a constant battle of wits between animals and zoo men. The animals insist on damming the running outlet of their pool. There is no difficulty, of course, if no leafy branches, or browse, are supplied and no mud is available. This method has been successfully used in the Detroit Zoological Park where a mother and kits in a moated, concrete inclosure became so accustomed to artificial feeding that they would emerge from their small brick lodge at any time at their keeper's call.

When closer approximation to natural conditions is desired, browse must be furnished, not only to supply food but to provide branches to use in conjunction with mud for building activities. Some means must then be found to prevent the beavers from flooding the area by raising the dam level. It is useless to pull the dam down to a suitable level every day, since an active colony will rebuild it every night, probably even higher than before. Something can be gained by shutting off or diverting the water supply in the late afternoon, but this affects the freshness of the water, and a heavy night rain could wipe out the gain.

It *is* possible to outwit the beaver, however. We did it for years by installing a standpipe outlet, set for the desired water depth and attached to a drainpipe that discharged outside

These beavers were part of a colony in the New York Zoological Park, photographed on a rare occasion when they fed on land in daylight.

the enclosure. Installed about 25 feet upstream from the dam, it was surrounded by a wire-mesh cylinder 3 feet in diameter that permitted free flow of water but effectively blocked all efforts to close it off. There is an alternative method by which a pipe, its upstream end inserted in a wire-mesh cylinder, is thrust through the dam and extended outside the enclosure. Either device keeps the water from flowing over the dam and thus there is no incentive for the beavers to raise the level of the pool.

In our own situation, we won the battles but lost the war. Beavers are nocturnal animals and they were seldom seen in our beaver pond during full daylight. Activity might begin about 4:00 P.M., an hour before the zoological park's closing, although on heavily overcast days an occasional specimen might be seen abroad at almost any hour. By and large, beavers under natural conditions were not a satisfactory exhibit and when, finally, the outlet area of the drainpipe rose to a level that prevented runoff, because of an accumulation of silt that could not be removed, the beavers soon succeeded in flooding the inclosure to a point that led to the dissolution of the colony. Now the beavers have been gone for many years, and each spring Canada geese nest on the small island that once was an anchorage point for the beaver lodge.

THE WOLVES, WILD DOGS, AND FOXES

Some animals are so firmly embedded in legend and fable that they occupy a disproportionately large place in the zoo visitor's mind. The wolf is certainly one of these beasts of fearful imaginings; in the ideal zoo, a pack of wolves should always be skulking through a darkening wood, and preferably slavering as they skulk!

The reality of actual exhibition is usually quite different, however. In fact, I recall only two really excellent installations. One is a well-done concrete representation of a rock niche at Skansen in Stockholm. When I saw it last, this inclosure contained a pair of European wolves with a litter of puppies 2 or 3 weeks old, plainly visible beneath an overhanging ledge. This seemed to me an ideal arrangement for a small family unit. On a more imposing basis is the superb wolf wood at the Whipsnade establishment of the Zoological Society of London. Here a pack of wolves is shown in large twin inclosures, one of which is always out of use. They are thickly planted with straight pines and floored with thick deposits of dry needles. The animals are transferred periodically from one section to the other for sanitary reasons. There is sufficient space to avoid serious fighting, and the glimpses of wolves skulking among the tree trunks, in the heavy shade, are just what the romantic imagination demands. I must say that I never observed them slavering, however.

One drawback to wolves is that they breed readily in confinement—sometimes too readily—and consequently many American zoological gardens keep small groups of one sex only; even dog wolves that have been reared together or otherwise become familiar with each other get on perfectly well if there are no females about. The reason for this segregation is that surplus wolf pups are not readily disposable and in quarters roomy enough for a pair but no more, the father will not tolerate young males for long after weaning.

As might be expected, the longevity of wolves is about the same as that of the domestic dog; we kept a Great Plains wolf (received as a pup) for 15 years, 7 months, 10 days. The best record is probably that of a wolf of the same race that lived in the National Zoological Park in Washington for 16 years, 3 months, 5 days.

The coyote shares with the wolf the interest of Americans, especially those who remember its prominence in cowboy ballads and western songs. Consequently, most zoological gardens exhibit them, although they do not meet with great favor—again largely because of the difficulty of disposing of the surplus. When one pair may have as many as 11 pups at a birth, the size of the problem can be imagined.

The dingo of Australia is an animal of particular interest and there has been much speculation about its existence in the land of marsupials. It is generally agreed that the dingo must have been brought to Australia as the companion of very early human immigrants, presumably from the Malayan region. It is supposed that it eventually reverted to a wild state, finally becoming established over most of Australia. After the country was settled by white men, the dingo proved to be a scourge on the great sheep stations of the interior, and bounties were offered for its destruction. So many were destroyed by professional "doggers" that the true dingo has now become scarce except in the most remote areas.

A rather tall, rangy animal with prick ears, a somewhat bushy tail, and coarse hair of yellowish brown, the dingo is well known to zoo visitors everywhere. Early writers described the dingo as being red or black and marked with white, and it is true today that many zoo specimens have white legs, tail tips, or even facial blazes, and are further disfigured by hanging ears and a tail curled over the back. Perhaps a plea for an upstanding solid-red dog with strongly

The maned wolf of South America has been aptly called "a fox on stilts."

erect ears and a thickly haired tail carried well down is a plea for a false image. But at least such an animal, even with a touch of white on a toe or two, is not an eyesore.

We can only make suppositions about the origin of the dingo, and the same is true of the domestic dog. Much has been written concerning its possible ancestors, but knowledge has advanced little beyond the conclusions that it had a multiple origin from various canid ancestors, including the wolf, the jackal, and other locally available species such as the coyote in North America. All these animals are known to have produced hybrids, usually fertile, with domestic dogs. The dingo has a place in the picture, too, for not only does it represent a primitive type but also it appears to have in-

In earlier days the Indians of South America professed
great fear of the bush dog, known
in British Guiana as the "Warracaba tiger."
Captive specimens have been anything but fearsome.

fluenced the development of Australian breeds as well as those of domestic dogs of nearby islands. The village dogs of southeastern New Guinea are black or yellow, self-colored or pied with white. Although rather slim, they otherwise resemble dingos and, like them, do not bark. I have a clear recollection of an old, solid-yellow dog, grown paunchy with age and with a definite aversion to white men, that would pass anywhere as a dingo.

Any zoo that had the space and the interest—and the luck —to create even a fairly comprehensive exhibit of the wild dogs of the world would have an undeniable attraction, for the wild dogs are almost as varied naturally as domestic dogs are by selective breeding. Such an exhibit would include the long-legged maned wolf of South America, the raccoon-like dog of Asia with its black facial mask, the descriptively named crab-eating dog of tropical South America, the savage dholes of Asia, the feared hunting dogs of Africa, and many others. Many of them come into captivity so rarely, however, that it would be difficult to form anything like a complete collection.

We have, for instance, been able to exhibit the South American bush dog only twice. This is the famous "Warracaba tiger" of which so many stories used to be told in British Guiana. I have heard many of them, although I never encountered an Indian who had actually seen the "tiger." The older Indians, deep in the bush, were undeniably afraid of it, whether they had seen it or not. It was, then, something of a shock when a bush dog was sent to the Bronx Zoo by our animal collector, and it turned out to be a heavy-bodied, stumpy-tailed, short-legged little creature that certainly looked as if it could bite savagely, but that in practice was more inclined to come to the front of the cage, sniff at a proffered hand, and wag what little tail it possessed. We were surprised to find that it could climb a 2-inch-mesh wire netting with ease, so that the top of its cage had to be covered. The Indians in British Guiana claimed to recognize the Warracaba tiger by its voice, but the only sounds we ever heard it make was a continuous whine in a thin, high-pitched voice—certainly not a sound to strike terror into the heart of anyone.

More often seen in zoos is the Cape hunting dog once found over most of Africa, except for the desert areas of the north and the heavily forested regions of the west. Because of its serious depredations on domestic stock, it has been extirpated or at least greatly reduced in numbers in the more settled parts of its former range.

It is, to say the least, not a handsome dog; the general coloration is in blotches of black, gray, white, and yellow, so greatly varied that two animals exactly alike could hardly be found. A swift, relentless hunter, usually running in packs, the hunting dog pulls down and devours the largest antelopes and is execrated by sportsmen and farmers wherever it still exists.

In recent years, births and the successful rearing of hunting

Packs of Cape hunting dogs are serious scourges to domestic stock in Africa, but their pups are very much like puppies of any kind.

dogs have occurred so frequently in this country that surplus animals have actually become almost indisposable. This was not always so, however.

It was formerly our practice, when the birth of hunting dogs was imminent, to remove the male from the female's quarters. This is common practice where many animals are concerned, under the usually close conditions of captivity, and in any event we had the authority of another zoo man that the males "have no obligations whatever during the birth of the young." Late in November, 1959, it was obvious that our female hunting dog was pregnant, and in accordance with previous practice a large, covered nesting box with a single entrance was installed, and the male was removed to the adjoining cage. When the first birth occurred, the female did not lick the pup but took it in her mouth and leaped excitedly against the partition, solid below and wired above, that separated her from her mate. After perhaps an hour the male was admitted and went directly to the pup that, already dead, had been placed on the floor by the mother. He licked and mouthed it but finally abandoned it. Five more young were born at intervals on the bare floor of the cage. Successive puppies received no attention from the mother, but each was carefully cleaned and gently mouthed by the father. It was then picked up and carried into the nest box by the mother. The birth completed, she then showed intolerance of the male and he was run into the adjoining cage, but with the door left open. In following days the female sometimes visited him there, but he was obviously not welcome in her compartment, and was never allowed to enter the nest box. On the morning of the day when the puppies were 14 days old, the eyes of all were found to have opened. On that day one stumbled out of the box. The male rushed into the cage, presumably bent on rescue, but this intrusion so disturbed the mother that she drove him back into his cage, the door of which was then closed. As long as this arrangement continued the animals remained calm, but if the male was moved while his cage was serviced, both became agitated, calling to each other in their strange, hooting voices. By the time they were two months old, the puppies were completely weaned, separated from the mother, wormed, and immunized. The parents were then reunited—and we had learned that the male *does* have an obligation during the birth of the young.

THE BEARS

So homogeneous are the bears that any given species is instantly recognized as a member of the family. Their obvious distinguishing characteristics—heavy bodies, short ears, plantigrade feet, and abbreviated tails—apply also to the giant panda so aptly that early naturalists knew this strange creature as the parti-colored bear. But aside from this species, which is more nearly allied with the raccoons than with the bears, no mammal is likely to be confused with the bear, and no bear is likely to be mistaken for something else.

Bears are well distributed in North America, Europe, and

*Smallest of all the bears is the Malayan sun bear,
four feet or less in length and reaching a
weight of scarcely one hundred pounds.*

Asia. One species, the spectacled, is found in northwestern
South America, and even Africa seems once to have been
represented by the now presumably extinct Crowther's bear
of the Atlas Mountains.

Much alike as bears are in general characters, there is great
variation in size, ranging from the huge brown bears, the
world's largest terrestrial flesh-eaters, to the diminutive sun
bears of the Malaysian region. Almost all are omnivorous,
eating flesh when it can be had, otherwise resorting to fruits,
berries, roots, leaves, and even grass. A special fondness for
honey seems to be a common trait.

Except for the male polar bear, which remains active during the months of deep cold and darkness, bears of the north
usually pass a period of dormancy in a sheltered nook during

the winter months. Males and unmated females occasionally arouse from sleep and wander about, while mothers with newborn young drowsily nurse their helpless offspring. The dormancy of bears is not the deep torpidity of true hibernation, so that they may not be disturbed with impunity.

Since the days when bears were used in the gory exhibitions of the Roman arenas, their treatment has progressively bettered through the centuries. The famous Bear Pit of Berne, dating at least from 1480, was occupied, off and on, up to 1825, another and larger pit coming into use in 1857.

When construction of the bear dens in the New York Zoological Park was begun on September 7, 1898, a new era was heralded. Bears were no longer to be housed in pits but would have the run of ample inclosures, fully exposed to fresh air and sunlight, with rocks for climbing, deep pools for bathing, and snug shelters against severe weather. Skeletons of old cedars, dried and seasoned, were set upright in beds of concrete for the benefit of the more agile species. Doors of steel confined the animals to their dens for the daily cleaning of runs and pools. Steel bars 9 feet high and surmounted by inside overhangs were supposedly bear-proof and did constitute advanced design at that period. Later, when skillful climbers like American black and sloth bears found means for circumventing the overhangs and were able to make nightly forays in the grounds, returning at dawn, great sheets of metal ended their acitvities but added nothing of beauty.

During the period through which these once superb bear dens functioned, great advances in ideas for the maintenance of this group were made. These stemmed from the innovations of Carl Hagenbeck, developed early in the present century and quickly given worldwide approval. Great mountains of concrete, simulating rock, which concealed dens and passages that opened into runs divided by the same construction and separated from the public by broad open moats, soon became central features of many of the great zoological gardens. At the present time there seems to be a trend in newer construction toward detached, single inclosures as opposed to extended ranges.

In most American constructions, widths and depths of bear moats for the larger species are considered to be safe at 14 feet. At least one lesser figure is known to be unsafe: 10 feet, 6 inches. In an installation in which this was the width of the moat, a brown bear not fully grown succeeded in escaping— on opening day—by simply extending itself across the space. After the moat width had been extended to 14 feet, there was no further difficulty.

Because of their particularly variable temperaments, bears should never be fully trusted by keepers, and no den or inclosure containing animals should ever be entered.

In this country the black bear is the species most commonly kept in captivity. Since it is an especially skillful climber, extra precautions must be taken to prevent its escape; barred dens should be topped and possible clawholds carefully avoided in rock or concrete barriers.

There is a surprisingly wide variation in the size and weight of black bears. Probably 200–300 pounds would represent a fair average, but much heavier individuals have been recorded. For instance, a male from Anticosti Island, Quebec, weighed 635 pounds, and specimens from Louisiana and Yosemite, 671 and 680 pounds, respectively.

It is well known that bear cubs are very small at birth, but just how small is revealed by weights recorded at the Copenhagen Zoological Garden where the heaviest of four black bear cubs was approximately 12.7 ounces and the lightest about 9.3 ounces. Such tiny creatures cannot, of course, fend for themselves; they are born while the mother is dormant and remain in seclusion with her until they attain size and strength. In our only experience with black bear births, the cubs left the den when they were 2½ months old and a few days later were seen nibbling on scraps of meat.

Our best longevity for a black bear is 19 years. The Frankfurt Zoological Gardens, however, kept one for 25 years, 11 months, 6 days, at which time it was killed by another bear.

The great brown and grizzly bears include the largest and most powerful of the living carnivores. Because of their great size and potential ferocity they are well represented in folk-

*Some Alaska brown bears may achieve very
great weights. One Kodiak bear,
in captivity, weighed 1,670 pounds.*

lore and have been considered as among the most desirable
of sportsmen's trophies. Both "brownies" and grizzlies, when
taken young, accept captivity well and are kept by most
zoological gardens, although they are certainly not inexpen-
sive exhibits. The amount of food required to keep an adult
animal in good condition is formidable; the daily provision
we make is 10 pounds of raw horse meat, 5 pounds of whole
fish, and 5 loaves of bread, with apples, vegetables, and
greens as available.

The European brown bear reproduces so freely in captivity that disposal of the offspring is sometimes a problem.

Some Alaska brown bears may achieve very great weights. One commonly quoted is 1,656 pounds for a specimen from Kodiak Island. Details about a very large male Kodiak bear were furnished me by the secretary-treasurer of the Cheyenne Mountain Zoo in Colorado Springs, Colorado. The bear was received as a cub, directly from Kodiak Island, and when it died some 15 years later it was transported by means of a derrick and a truck to the scales of the nearby Broadmoor Hotel, where it was found to weigh 1,670 pounds.

As might be expected, cubs of the brown and grizzly bears are somewhat heavier at birth than those of the smaller black bears, but they are still minute in proportion to the bulk of the mother. For American brown bears the weight of cubs is usually given as about 1½ pounds and the length as 8-9 inches. European brown bear cubs seem to be smaller; figures of 9-13 ounces are recorded.

It is a rather curious fact that although breeding successes with grizzly and Alaskan brown bears have been few, the

A grizzly bear cub can be entertaining and appealing but as an adult it is among the largest and most powerful of the world's carnivores.

European brown bear reproduces so freely and the mothers rear their young so well that disposal has sometimes become a problem, as with American black bears. Perhaps this is because the European form has a longer captivity history; captivity-bred specimens are readily available and they should be expected to be more amenable than animals bred in nature, even though the latter are subsequently hand-reared.

Compared with most other bears, the browns and grizzlies live a very long time in captivity. We kept a Peninsula brown bear for 36 years, 10 months, 6 days; the Baltimore Zoo records 33 years, 8 months, 7 days for a grizzly; and Skansen in Stockholm had a European brown bear for the astonishingly long time of 47 years.

Closest in size to the big browns among bears is the male polar bear—the male being considerably larger than the female. A male killed in Hudson Strait is said to have weighed more than 1,600 pounds, and although this weight is generally accepted, it is certainly far above the average. Nine males killed in the wild averaged 900 pounds and the average of seven females was 700 pounds.

In captivity no bear is more popular or more sought after than the polar. Its unique white pelage, great size, and skillful water play make it most attractive. The pool in our present polar bear inclosure is roughly circular and about 20 feet in diameter, with the water 7 feet deep. As the bears plunge through the floating ice cakes in winter, visitors shiver in delighted astonishment. The bears' obvious enjoyment of bathing in hot weather arouses both approval and envy. Like other bathers, they enjoy water toys and will play endlessly with a small aluminum keg.

Unfortunately, the temper of the polar bear is more unpredictable than that of any other; thus it is especially important that visitors, made incautious through admiration, be prevented from getting too close to the exhibit area. For many years we have posted signs around our polar bear inclosure saying in stern bold type: "Keep Back, or You May Be the Next to Be Hurt." As far as I know, there has never been an antecedent to that ominous "Next," but the warning is salutary. I do recall, however, one occasion when a visitor came very close to being the "next to be hurt." It was a young boy who, delighted with the antics of three polar bears in their pool, determined to take a photograph. He climbed over the guardrail—directly under the warning sign, as it happened—and sighted his camera through the bars at one of the bears that was attempting to juggle the slippery aluminum keg. Intent on focusing, he did not realize that one of the animals had quietly left the pool and was approaching him silently on hairy-soled feet. The camera suddenly disappeared from the boy's hands as the bear made a swipe with his claws and swept the camera through the bars and into the water. The boy fell back in terror and escaped a second lethal swipe.

Polar bears are frequently born in captivity but are seldom reared. At birth they are covered with fine white hair. One measured, at birth, 10¼ inches from the tip of the nose to the base of the tail and weighed 1 pound, 3 ounces; another was only 9½ inches long but weighed 1 pound, 10¾ ounces. The eyes were closed. For reasons that are not hard to understand —by anyone who has ever seen a baby polar bear—cubs are enormously popular with zoo visitors, as the Zoological Gardens of London discovered in 1949 following the birth of Brumas, the first polar bear to be reared in captivity in Great Britain.

THE RACCOON FAMILY

Many zoological families are comprised of members that bear little superficial resemblance to each other. A prime example is the raccoon family that harbors not only the small, furtive, and nocturnal cacomistle but the comparatively huge and bearlike giant panda. It is not the externals, of course, that determine an animal's relationships, but its skeletal and other anatomical structures, and on this basis even the giant panda is closer to the cacomistles, raccoons, coatis, kinkajous, and olingos than it is to the bears.

The raccoon and the giant panda are no doubt the best known members of the family, but all the others have their points of interest. Westerners in particular are familiar with the cacomistle, sometimes known as the ringtail or civet cat, that is found from southern Oregon to Mexico. With its grayish body, black facial markings, and thick tail strongly barred with black, the cacomistle might almost pass for a diminutive raccoon except for the absence of the black mask across the eyes. Being a night-active animal, and usually a wary one—

although there are instances of its becoming tame enough to feed from backyard feeding stations in California—it is seldom seen, although it may be locally abundant. In spite of its particularly innocent expression, the cacomistle is a fierce and relentless hunter, preying extensively on small mammals, birds, reptiles, and insects, but also taking a variety of vegetable food.

Any inclosure meant to restrain a cacomistle must be tight indeed, for this little animal can run, leap, or ferret its way out if the slightest opportunity appears. I can recall one that managed to escape, and during a single night it destroyed an entire breeding colony of mourning doves in an aviary several hundred yards away across a large pond and supposedly fully protected from such marauders.

There are few actual breeding records for the species in captivity, but I can report a near miss in the Bronx Zoo. One June morning a female of a pair kept caged indoors gave birth to three young. She had chosen a rock crevice for their nest and cared for the young well, without interference by the male. Three mornings later, however, the cage was found in wild confusion, the tiny, sightless, fluff-covered young dead, the parents dishevelled, and on the floor the mangled remains of a very large rat buried under the dry leaves. How this intruder had managed to get in is still a mystery, but the loss of the young undoubtedly resulted from the ensuing struggle.

In spite of its natural shyness, the cacomistle accepts captivity readily enough. We have managed to keep one specimen for as long as 14 years, 3 months, 11 days.

Far more familiar than the cacomistle to most people is the raccoon, which is found from southern Canada over most of the United States and southward through Mexico and Central America to Panama. In spite of its nocturnal habits, so ubiquitous is this attractive animal that few of the world's zoological gardens have missed including it in their collections, and those in North America, at least, have had to contend constantly with the problem of oversupply. Hand-reared raccoons make especially attractive pets, but unless kept under close surveillance and control are likely to get out of hand as they mature. It is then that the owner seeks to transfer his problem to the local zoo, which cannot always refuse to accept the donation.

In nature the raccoon eats practically any sort of animal or vegetable food that it can secure. Small mammals, birds, reptiles, fish, crayfish, and insects, as well as nuts, fruits, and cultivated vegetables—all have a part in its menu. The habit of sometimes immersing its food in water before eating it is well known—indeed, the raccoon's scientific name, *lotor*, means "the washer," and its German name, *Waschbär*, literally means "wash bear."

The raccoon is as much at home in the trees as on the ground and is likely to spend the day curled up in a crotch or cavity. It is a common animal in the wooded sections of Westchester County immediately north of New York City, as many a suburban gardener knows. I have even seen a raccoon, caught by daylight when perhaps it was returning home from a raid on a vegetable garden, asleep in midmorning high in a tree between a major access parkway to New York and a suburban home 200 yards away.

Its tree-climbing ability poses problems for zoos. Any sort of natural-appearing inclosure must have smooth walls and an overhang at the top, which is of course easy enough to accomplish. If trees are provided as an outlet for the animal's desire for climbing, it is necessary to make sure they are not too close to trees outside the inclosure, for otherwise there is sure to be a mass exodus of raccoons. We met the problem in the Bronx Zoo by encasing the trunks of trees in the raccoon inclosure in tin "pants" to a height of about 8 feet. This, we were confident, would take care of the escape problem offered by the live trees in the inclosure; and for climbing purposes we offered dead and well-branched trunks carefully set in concrete at a safe distance from the live trees. What we failed to take into consideration, however, was the possibility that wild raccoons from the outside might attempt to get into our inclosure, which had certain undeniable attractions such as free food and 8 female raccoons. We were awakened to the situation one spring when we discovered that our in-closure contained 9, instead of 8, raccoons, and that several of our females were pregnant. A wild male, presumably wandering down the Bronx River that flows through the zoo, had invaded the inclosure by climbing live trees outside, connecting with the live trees inside, and sliding down the protective iron pants.

When the breeding of raccoons is desired, this is easily accomplished. In the latitude of New York, mating usually occurs during February and March, followed by a gestation period of 63 days. The young, from 1 to 6 in number, are sightless and very lightly furred but are able to leave the den with their mother when they are 6 to 8 weeks old—perfect miniature replicas of their parent. If a suitable nesting den has been provided, the mother is usually able to defend the young against any designs of the father, but if quarters are small he is best removed. Once they are out of the nest, the father usually accepts the young or at worse is indifferent. Oddly enough, raccoons seem to be prone to color variation and both albinistic and melanistic—white and black—babies are not uncommon.

It has always been a little hard for me to understand the charm, as a pet, of another member of the raccoon family, the coati. My feeling about the animal is, I admit, colored by the many accounts I have heard over the years of pet coatis getting loose in their owner's home and all but completely wrecking the interior. There must be few small animals as innocently destructive as an exploring coati. Still, they are regularly offered for sale in pet stores and presumably find purchasers.

Coatis are forest animals, ranging from the southwestern United States through Mexico and Central America as far south as Paraguay. Color varies among the many forms from gray to brown or bright bay, but in all the tail is strongly ringed and is usually carried erect. The mobile nose is noticeably prolonged—a nose expressly formed, one might think, for poking into trouble!

Less nocturnal than the raccoon, the coatis prefer morning

and evening for their forays, often hunting in small bands perhaps composed of mother and young. They ascend trees in
search of fruit, as well as hunting over the ground below;
nothing is overlooked, whether it be small mammals, birds,
reptiles, insects, or fruit. Their sensitive noses are constantly
testing and add to the impression of implacable ferocity, well
confirmed by their nature.

The kinkajou, known popularly as honey bear, is another
member of the family sometimes sold as a pet. It shares with
the coati the more tropical portion of the latter's range, from
southern Mexico to central Brazil. Uniformly light brown,
honey-colored, with short, soft hair, a round head, short
muzzle, and comparatively large eyes, the kinkajou is undeniably an appealing little animal. The most characteristic
feature is its tail, which is long, tapering, and strongly prehensile.

Perhaps because of its nocturnal habits, the ways of the
kinkajou in nature are not well known. It lives almost entirely in the trees of the great forests, seldom descending to
the ground.

Most of the kinkajous seen in captivity have been reared
by hand or at least taken while very young and so make
gentle and confiding pets. This should not be taken as indicative of the basic nature of the animal, however, for the
occasional wild-caught adult is savage enough. Even a perfectly tame kinkajou, if sufficiently frightened or annoyed,
can inflict a painful bite.

The supposition is that in the wild kinkajous feed largely
on fruits, and probably insects and small birds. This is borne
out by their diet in the zoo—oranges, apples, bananas, and
grapes, with small allowances of bread, carrots, and peanuts,
which they especially like. They also take a mixture of raw or
cooked chopped meat, dog biscuit, cooked or raw egg, bone
meal, and cod-liver oil, the lot sweetened with condensed
milk. So avid are they for ice cream, which is drawn into the
mouth with the extensile tongue, that one of the keepers in
our small mammal house some years ago worked up a little
act for the entertainment of visiting dignitaries. He would

offer an ice-cream cone to a kinkajou and as soon as the animal seized it and began licking it, the keeper would swing the animal up by its tail and bring the grasping tip in contact with a horizontal bar in the cage. Invariably the kinkajou would grasp the bar with its tail and hang upside-down while continuing to lick the melting ice cream as fast as its tongue could flick in and out. "And he never spills a drop!" the keeper would triumphantly conclude his demonstration.

Despite the frequency with which they are kept in captivity, they are not often induced to breed. We have done quite well with breeding, however, since exhibiting the animals under red light to reverse their night activity. The first birth, under these circumstances, was to one of four females kept with a male. In the general activity of the group, as we were able to observe it for the first time under these lighting conditions, the young animal appeared in some danger, for although none of the occupants seemed to threaten it, neither did the mother offer it protection. Eventually, however, we were able to determine which of the females actually was the mother, and the others were removed. When a section of a hollow log about 18 inches high was stood on end in the cage, the mother seized the baby in her mouth by its neck and deposited it in the log. Here she cared for it well and eventually reared it.

Kinkajous are long-lived in captivity, and several have lived in the Bronx Zoo more than 10 years. We had high hopes that Jimmy, the animal who was many times the star in our keeper's ice-cream cone demonstration, would set a world's record for longevity. He died, however, when he had been with us 22 years, 1 month, 18 days, and at least two other kinkajous have done somewhat better. The record seems to be held by an animal in the Zoological Gardens of Amsterdam—23 years, 7 months.

Closely allied to the kinkajou and apparently often confused with it is the olingo, which ranges from Nicaragua to northwestern South America. Few specimens seem to have found their way into captivity, although the Zoological Gardens of London exhibited one as long ago as 1894. An olingo received at the National Zoological Park in Washington in 1956 was thought probably to be the first exhibited in this country. It does seem possible that the animal is less rare than has been supposed, for its resemblance to the kinkajou certainly leads to confusion. In fact, our first olingo, also received in 1956, was obtained from a pet-seeker who had bought it as a kinkajou and was disappointed by its unfriendly nature.

Seeing the two animals side by side, it is not particularly difficult to distinguish between the kinkajou and the olingo. The latter is a grayer brown than most kinkajous, with pale gray face and noticeably longer and more pointed muzzle. Its most obvious characteristic is the long tail, which is very faintly ringed, somewhat bushy, and—surprisingly—not prehensile. The olingo is far more active than the kinkajou, even by day, and runs and leaps with an agility seldom shown by its relative.

We know even less about the habits of the olingo in the wild than we do of the kinkajou's. The statement that it is "largely arboreal, nocturnal, and frugivorous" may well have its origin in the report of one naturalist who shot both kinkajous and olingos from the same tree in Panama, where they had been feeding on fruit at night!

Two related animals as unlike, externally, as the lesser panda and the giant panda are hard to imagine. The lesser panda does, however, suggest a raccoon, and a particularly attractive one since it has a full coat of bright chestnut, a light-colored face, black legs and abdomen, and a bushy, sharply-ringed tail. There are two races, one from the eastern Himalayas in Nepal and Sikkim and a slightly larger one in the mountains of southeastern China and the borders of northern Burma.

Our first lesser panda, quite possibly the first to be seen in this country, came in 1911. In those early days lesser pandas were kept in our small mammal house and were coddled as rarities are likely to be, despite which—or because of which—they could not be made to thrive. Later we began keeping them in a walled and moated outdoor inclosure sporadically used for raccoons, with much better results. We found that they are essentially day-active, although inclined to sleep a good deal through the middle of the day, becoming most active in morning and evening. Also, they are not at their best on the ground and are thoroughly at home in the trees. Even adults readily accommodate themselves to changed conditions when they come into captivity, and we have found them to be gentle and non-aggressive and willing to allow some familiarity on the part of their keeper, although, if actually touched, they usually show their resentment by emitting explosive, coughing sounds.

At liberty in the outdoor inclosure, the animals spent much time in their climbing tree, a dead cedar from which the bark had been stripped. It was in this tree that the only offensive gesture we have seen occurred; one male crowded another from the tip of a branch, so that he fell to the ground and was killed. They showed a definite distaste for wetting themselves, and when, apparently by accident, one got too far into the pool, it hastily scrambled out and shook itself with vigor, lifting its feet gingerly like a cat. Coming from high mountainous regions, the animal is, of course, indifferent to cold and here, at least, it has shown no inclination to winter dormancy.

In the giant panda the raccoon family reaches its culmination, at least as far as the zoo-going public is concerned, for I cannot think of any animal better formed by nature to attract and delight everyone, child and adult alike. Since 1936 there have been nine giant pandas in the United States, four in the Bronx Zoo, three in Chicago's Brookfield Zoo, and two in the St. Louis Zoo, but unfortunately it may be a long time before they again have a chance to amuse and instruct Americans. Found only in the province of Szechuan in western China, directly north of Yunnan, where they live in the dense thickets of bamboo that cover the steep mountain slopes at elevations of 5,000 to 10,000 feet, the animals are inaccessible to anyone except the Chinese, and as long as there is a government embargo on trading with the Chinese no giant panda can be brought into this country. The restriction applies even to specimens that may be acquired by dealers in countries that do have commercial dealings with China.

The giant panda is a true giant of its family. Adults may measure 6 feet in length, disregarding the stumpy tail, and may weigh more than 300 pounds. Thickly furred with coarse white hair, having legs, shoulder band, and eye patches of deep black, the animal is a roly-poly giant, seemingly boneless in its wriggling contortions, and endlessly playful.

Although it was made known to the outside world as long ago as 1869 by the French missionary, Père David, it was not until the end of 1936 that the first living giant panda to leave China was brought to New York by Mrs. Ruth Harkness. Su-Lin, as the animal was named, was still an infant, fat, pudgy, and unable to walk with ease, and he spent most of his time lying happily on his back. Early the following year he was acquired by the Chicago Zoological Park where he was, of course, an instant and great attraction. Panda dolls, from miniatures to realistic representations more than life size, were soon on sale all over the country. Su-Lin's untimely death in 1938 was felt as a personal tragedy by many people. Before his death, however, Mrs. Harkness had brought him a "mate"—which proved also to be a male. The third and last of the Chicago pandas was Mei-lan, which came in 1939 and achieved fame by living until 1953. This span of 13 years, 10 months, appears to be the greatest so far accomplished by a captive giant panda. The St. Louis Zoological Park acquired its two giant pandas in 1939 and one of them lived for more than 12 years, just about a year short of Mei-lan's record.

The felicitously named Pandora was the first of the Bronx Zoo's four pandas. She came in the late spring of 1938 when she was still a baby, weighing 35 pounds, and proved to be so friendly and amenable that she was undoubtedly one of the factors that led to the decision by New York Zoological Society to construct an exhibition building at the New York World's Fair of 1939 and 1940. Pandora was the star attraction of the exhibit, and it was hard to avoid the conclusion that she enjoyed entertaining her thousands of visitors by turning spineless somersaults, wearing her food pan as a hat

*A giant panda certainly does not perform its antics
with the knowledge that it is being amusing,
but it is true that almost everything
the animal does is entertaining.*

or trying to fit her broad rump on a tiny block of wood.

When Pandora' life ended in the spring of 1941, from an undeterminable cause, she weighed 230 pounds. She had reached a maximum of 265 pounds in the previous October.

Our experience with a male giant panda, Pan, who came in 1939, was not as happy as that with Pandora. For some reason, Pan was unfriendly, even surly, in his reactions to his keepers. He lived only a year and never adapted fully to captivity conditions.

Our last giant pandas, Pan-dee and Pan-dah, named through the sometimes unfortunate medium of a public contest, were received as gifts of Mme. Chiang Kai-shek and her sister, Mme. H. H. Kung, and an official of the zoological park went to China to receive them and bring them to America. After an exciting wartime journey by airplane and steamer beginning three weeks before Pearl Harbor, the animals were finally landed safely in New York on December 30, 1941.

A newly built inclosure was ready for them, 60 by 75 feet, surrounded by a stippled concrete wall 8 feet high at its lowest point. A huge, smooth, glacier-scored outcrop of rock occupied one corner, and a pool 15 feet across lay near the center, shaded by a tall hickory tree. Several slender cedars, set in openings in the natural rock and tile floor, enhanced the effect of a Chinese garden.

Since the pandas had come through warm climates only to reach New York in midwinter, they were kept confined in mildly heated indoor quarters at the rear of the inclosure for the remainder of the cold season, but they were given their liberty outside on the increasingly frequent warm days. We soon learned, however, that giant pandas have no fear of the cold, and in subsequent years we made no effort to keep them inside in winter. Indeed, they obviously enjoyed frolicking in the snow, and one of the most entrancing motion-picture records of the two animals shows them emerging from their night-shelter house on a sparkling winter morning and turning rapid somersaults in the deep snow all over the inclosure.

Despite our previous experience with two giant pandas, we had much to learn. One lesson concerned their climbing ability, and it was given in an unforgettable fashion by Pandee on one of her earlier visits to the outer precincts. She climbed the tall hickory tree to a height of 40 feet, where she remained for 41 hours before she could be persuaded to come down. Since she had chosen a crotch perch from which she would have landed on the edge of the granite outcrop if she had fallen, the concern of the entire zoo staff during this period can be imagined.

Pan-dee was finally induced to descend by the curator of mammals, who resorted to psychology. Knowing that Pan-dee and Pan-dah each had a distinctive food container, from which they were accustomed to eat their morning meal of a kind of mush, he had Pan-dee's dish brought out and placed on the ground in full sight of the animal in the tree. Pan-dah had been deprived of food for some hours and was then released to feed from Pan-dee's dish. The meal had scarcely started when Pan-dee stirred and with sure-footed haste descended from the very top of the tree, ambled to the feeding dish, bumped Pan-dah out of the way, and began to feed from "his" dish. Needless to say, any more expeditions to the top of the hickory tree were quickly ended by encasing the trunk in metal "pants."

On their arrival and for some time thereafter, Pan-dee was considered to be a male and Pan-dah a female. Later it was determined that both were females.

The food of the giant panda in nature is reported to consist solely of leaves, stems, and even fairly heavy stalks of bamboo. For a time we had suitable bamboo shipped regularly to the zoological park from the southern states, but eventually we discovered a species hardy in the New York area that the animals would accept. An exclusively bamboo diet in the wild seems so specialized that the "civilized" diet our animals accepted—and thrived on—is a little surprising. Pandora, when she first came to us, took milk, egg, Pablum, honey, orange juice, fish oil, green cornstalks, willow sprigs, celery, lettuce, Swiss chard, beet tops, and baked potato. This willingness, however, may have been only the eagerness of a young animal to investigate strange offerings, for she soon refused to eat anything except bamboo and a semiliquid formula to which ground vegetables might be added. Neither Pandora nor any of the other giant pandas would take flesh in any form.

Our giant pandas always appeared to be uncomfortable during periods of greatest heat, but, unlike the lesser pandas, they had no particular aversion to water as a means of getting cool. When a fine spray was directed across the tile floor of the outdoor inclosure, they walked slowly to and fro through it. Standing water they seemed to consider just another obstacle and waded directly through it to reach an objective, in the usual panda fashion. When the weather was really hot, they often sat down in the pool for a brief soaking, although this always seemed to be an afterthought and not deliberate bathing.

Having a monopoly on the supply of giant pandas, and having many excellent zoological parks, the Chinese have been the first to breed the giant panda in captivity. The first of two births was that of a 5-ounce male, named Ming Ming, born to a female named Li Li, in the Peking Zoo on September 9, 1963.

THE MINKS

In the Adirondack Mountains of New York, I once was challenged by a mink for possession of a row of trout I had caught and laid on a log behind me. Only a sharp tap on the nose with the tip of my fly rod caused the animal to relinquish its claim, but its bold and forthright rapacity left an impression I have never forgotten and makes me regret all the more that the mink has seldom been considered as a suitable exhibition animal in zoological parks. Like their smaller relative, the weasel, minks are usually so high-strung that they do not thrive under conditions of bright light and the constant stir of visitors.

There are two species of mink—the American, which is found over most of the United States and Canada, and the slightly smaller European from central and northern Europe to Siberia. They divide their time impartially between land and water, hunting small mammals and birds in the forests and pursuing fish and other aquatic creatures in lakes and streams. And everywhere and at any time they are bold and

This domestic "sapphire" mink was captured near New York City by a fisherman who baited it within reach of his landing net by offering bits of food.

fearless, seemingly ready to face any potential enemy, however large.

The great demand for mink pelts by the world's fur trade once placed a serious strain on the natural supply, but the establishment of the American mink in captivity has relieved much of the pressure. By selective breeding a fantastic variety of mutational colors has been developed, and it was one of these mutants that for a time provided an interesting exhibit in our small mammal house. This was a "sapphire" mink, presumably a fugitive from a mink ranch, that was captured some 56 miles from New York City. According to the captor, who presented the animal to the zoological park, he first saw the animal when it came out of a clump of trees and approached the center of a fishing camp where hamburgers were being cooked on an open-air grill. The captor happened to have beside him a basket of chicken heads that he was using as bait for snapping turtles, and he tossed one to the animal. It bounded off into the woods with the delicacy and returned in about 20 minutes for another handout. On its third appearance the fisherman was ready with a landing net, which he threw over the mink as it came up to take food. Whether a truly wild mink would have allowed itself to be captured so easily is perhaps doubtful, but remembering the boldness of my trout-stealing mink in the Adirondacks, I have no doubt it would approach within the throw of a net.

THE WOLVERINE

Few mammals have a more unsavory reputation, much of it justified, than the wolverine. The broad, squat body may reach a weight of 50 pounds, perhaps even more, which, combined with immensely powerful jaws and the grim determination common to all members of the weasel family, makes it indeed a formidable adversary. Robber of trap lines and killer of prey many times its size, the wolverine thoroughly justifies the dread and dislike it inspires whenever it makes human contacts.

Given these characteristics, it is perhaps fortunate that the animal does not inhibit the more thickly populated parts of the New World or the Old. It is confined to the Arctic and Subarctic regions from Norway to Manchuria and from Alaska and Northern Canada to the borders of the United States, with southern extensions in the western mountains to

Keeper J. Coder holds a lapful of young wolverines, tame and engaging pets while they were small but eventually unmanageable.

Colorado and California. Actually, it is greatly reduced in numbers throughout its range and is close to extermination in the United States.

Our best experiences with wolverines came at a time when the large inclosure prepared for giant pandas was empty. At that time we were able to get a litter of four young wolverines of the European species that had been taken from a den in northern Finland and hand-reared. These young animals, a male and three females, were still completely tame and most engaging. They throve on whole cow's milk, which they drank by lapping, and a mixture of chopped raw horseflesh, dog meal, fine bone meal, and cod-liver oil. Their growth rate was surprising; weighing 8 pounds on arrival, the male was 38 pounds four months later, and the three females did almost as well. We were never able to get a continuing series of weights on them, for with increasing age and growth they became unhandleable. While they never actually attacked their keeper, they stood their ground so resolutely that it seemed expedient to work around them, rather than to test their tempers further.

As adults they alternated periods of activity with short naps by day and at night. While they adhered to the wolverine reputation for solitary living and seemed to have no attachment for each other, there was no serious quarreling. On the other hand, we found it impossible to introduce an adult female European wolverine a couple of years later; she was a stranger and was subjected to mild but such continuous hazing that eventually she had to be removed.

The reaction of the wolverines to the water of their pool and to a spray used during hot weather was curiously like that of the giant panda. They splashed doggedly through standing water or swam if they had to, treating it as a mere obstacle between them and an objective; they stalked stoically through the spray with apparently complete indifference.

The name "glutton" is often applied to the wolverine, but it seems hardly justified. The daily diet of one of our adult males, probably weighing in excess of 40 pounds, was only about 3 pounds of food. Ordinarily the wolverine does not take fruit, but on one occasion one that had refused all food for several days and was causing us great concern suddenly and unaccountably devoured a peeled banana. The next day it resumed its customary diet and rejected bananas and other fruit offered.

THE SKUNKS

It would be interesting to know how widely the misnomer polecat is still applied to the skunk in the United States. Certainly it was the common term in many parts of the country two or three generations ago. The real polecat is found over most of Europe and northern Asia but not in the New World, and since it is able to eject a foul-smelling fluid from the scent glands beneath the tail, it is readily understandable why early English-speaking settlers in America applied its name to the skunks they encountered in their new home. The skunks are New World animals, ranging from southern Canada to Patagonia, keeping to open or broken country and usually avoiding forests. Most widespread are the striped skunks that range from southern Canada across the United States to northern Mexico; then there are the attractive little spotted skunks found from southern British Columbia and southern Pennsylvania south to Costa Rica, and finally the beautiful white-backed hog-nosed skunk distributed from the extreme southwestern borders of the United States to Chile and Patagonia. They all differ in one way or another— but all are unmistakably skunks when angry or frightened.

All are patterned sharply in black and white, which is fair warning of their extraordinary ability to eject and even to direct the irritating, foul-smelling fluid for which they are famous. Confident of their ability to defend themselves, they are quiet, inoffensive creatures, going into action only as a last resort.

Their basically gentle nature has brought the skunks some popularity as pets, a situation that makes "de-skunking" imperative. Quick results may be had by mere severance of the ducts of the scent glands that lie at either side of the anus, but only complete removal of these glands can assure permanent effectiveness. When the subject has been anesthetized, a single incision will expose gland and duct. The latter is then tied off and the gland removed. The operation must be repeated, of course, on the opposite side. If ordinary antiseptic precautions are taken, there should be no adverse

*Most widespread of the skunks in North America is the
striped skunk. It makes an excellent pet
when it has been "descented."*

after-effects, although youngsters of 6 weeks of age seem
to make easier recoveries than the older skunks.

The really tame individuals, even though in full possession
of their powers, may never use them, even under pressure.
We once received a female striped skunk that had been a
pet of its donor for two years or more. It was a really beautiful
animal, docile and friendly, its coat full and well brushed and,
of course, completely odorless. The owner reported that
while the skunk had been descented, she had noticed that
when she playfully threatened it with a broom for some slight
infraction, a faint odor could sometimes be detected. This
animal had been a charming attraction in our children's zoo
for some months, played with and handled by hundreds of
children, when an attendant happened to remark that Pe-
tunia, as the animal was called, was giving off a slight aroma.
Our veterinarian was hastily summoned and his close examination revealed that the scent apparatus was intact—a condition quickly remedied.

Skunks are generally nocturnal and terrestrial in habit, leaving their dens in earth tunnels, rock piles, or hollow logs in the evening and returning at dawn. Their tastes in food are broad, including small mammals, birds and their eggs, reptiles, insects, and various fruits when in season. Even in the frigid north they do not truly hibernate in winter and, while they are likely to undergo short periods of dormancy, are often about searching for food in the milder periods.

As zoological garden exhibits, it must be said that tame, descented individuals, maintained as "keepers' pets," are most common. It is sometimes found that a pair or even several females and one male will live together reasonably well, but even then some quarreling is bound to occur. As winter approaches, the animals become sluggish and inordinately fat and do not fully regain their attractiveness as exhibits until spring comes again.

Anyone who has noticed the great numbers of skunks found dead on highways, struck by speeding cars, may suppose that their life span in the wild is very short. Even in captivity it is not extraordinarily long; we have kept the striped skunk for 6 years, 10 months, 14 days, and the National Zoological Park kept one for 7 years.

THE OTTERS

Many animals, especially juveniles, engage in play, but I know of none that excel the otters in sheer romping activity. Other animals may occasionally indulge in antics that seem to have little to do with their main pursuits, but otters seem to make a strenuous business of play for play's sake. This is true not only of the American otters, but also of the Eurasian species, the otters of the far north, and the giant otter of South America. Long, sinuous bodies, short legs, and broadly webbed feet are ideally adapted for the swift movement and the agile turns in water in which the river otters excel. They are said to be able to swim under water for a quarter of a mile, and two American otters were captured in one crab pot at a depth of 60 feet in Alaskan waters. No otter is at its best on land, yet river otters walk with undulating grace and gallop more awkwardly but at a fair speed and are able to make overland trips sometimes of considerable distances. Their habit of sliding down mudbanks into water or tobogganing on snow, apparently in a playful spirit, is well known.

Despite their engaging ways, otters are not beloved of fishermen, particularly in Europe. Their diet, however, is by no means confined to fish; it includes frogs, crayfish, insects, and any small mammals or birds they may be able to capture. In America they are sometimes accused of serious destruction of muskrats, and the Zoological Society of Ireland reported sadly some years ago that wild otters got into the waterfowl inclosure of the Dublin Zoo and decimated the collection.

The first essential of an otter exhibit is a pool of fresh, clean water, deep enough and long enough for swimming, diving, and play. Moderately running water is certainly preferable, if not essential. There should also be sufficient land area to allow reasonable space for the explorations in which otters seem to delight. For, while we tend to think of the otter in terms of water, the animal really spends much more time out of water than in. Another essential, of course, is an unclimbable fence or wall; one 4 feet, 6 inches high, will contain any river otter, provided snow does not raise the level inside it.

We have done well by our river otters. Their inclosure is approximately thirty feet in diameter, floored either in a natural outcrop of rock or asphalt and concrete to prevent the persistent digging that is characteristic of them. A trickle of running water crosses the area in front of their shelter containing individual dens, finally ending in a pool about 15 feet long, 4 feet wide, and 2½ feet deep. Just behind the pool is a natural rock 3 feet high, on the face of which we built a "slide" of steel-troweled concrete, made slippery by a flowing film of water, intended to convey playful otters to the pool itself. Only once, as far as I know, has this slide been used for its intended purpose, and that quite by accident. On that occasion an otter carelessly attempting to walk across it missed its footing on the glassy surface and slid ignomin-

Otters have a great deal of curiosity about everything going on around them and often interrupt their play in the water to stand up and stare at visitors.

iously into the water on its back. Captive otters will slide freely in snow, but I have never seen one make proper use of the artificial slides customarily provided.

Since otters are usually considered unable to climb trees, we once were greatly surprised to find a female Florida otter in the habit of ascending to a height of 6 feet or more in a small cherry tree, which leaned slightly, in the inclosure. She accomplished this by hugging the trunk with her forelegs and inching herself upward. This she did easily enough, but when she attempted to descend by reversing the process, she was awkward and uncertain. A final fall to the concrete floor fortunately caused her no injury but it led to providing the tree with metal "pants."

For animals rarely exceeding 25 pounds in weight and often considerably less, otters are heavy feeders. Our river otters have a daily allowance of 2 pounds of butterfish, a quarter-pound of smelts, a half-pound of whole raw meat, and a half-pound of the chopped raw meat, dog meal, bone meal, and cod-liver oil mixture.

Our experience in breeding otters is limited to the activities of a pair of Florida otters. Late in December the female began working on a well-constructed nest of small branches, dry leaves, and straw in one of the compartments of the shelter. We suspected a birth to be imminent, but one morning an ambitious relief keeper removed the nest material as unsightly—and the next morning two female cubs were found dead on the wire mesh of the compartment floor, evidently born during the night.

Three years later, this time in January, we had better luck. First notice of the birth was the ejection of the male from the shelter by the female. During the first day he reentered several times but always emerged, hurriedly, after a brief period of excited squealing. He finally gave up the attempt and sought refuge in a small wooden hutch placed in the inclosure for his use. The female's food was placed outside the entrance to her shelter, and when she came out to eat, the male remained discreetly in his box. A set ritual soon developed: when the female was about, the male remained hidden and came out only when she was tending her cubs. If he happened to be outside or even in the pool, the slightest sound from the nursery sent him scuttling for shelter. Later, when the mother began coming down to the pool to bathe, she did not disturb the male as long as he kept to his box but drove him to it furiously if she chanced to find him outside it. Obviously, he should have been removed entirely, had other quarters been available.

The first youngster to be seen outside the shelter came out alone when it was 38 days old. It wandered nearby for a minute or two, when the mother rushed out, threatened the male, which had looked out from his box, seized the cub by the neck, and dragged it, walking backward, into the shelter.

Such excursions soon became daily events and all four cubs, always accompanied by the mother, often ventured several feet from the shelter. At the age of 63 days they were first seen eating the chopped meat and dog-meal mixture and drinking water from a tiny flow. Two weeks later they were seen playing together in the flowing water, which we had cautiously reduced to the depth of 1 inch. When they were just short of 3 months old they were frolicking in the deepest water they could find, about 4 inches, at the neck of the pool. We decided to double the depth. Ten days later the outlet became blocked during the night and the next morning mother and brood were found swimming and diving in water 14 inches deep. At that point we decided that the mother knew what she was doing, and restored the full depth of 2½ feet.

During this entire period the male seemed much interested in the young, but the female never allowed him to approach them. When one of the wandering young came near his shelter, he cowered within it. This resistance by the female gradually broke down, however, as she busied herself carrying food to her active offspring, and the male was permitted more freedom. One day the father was seen to pick up a fish near a feeding youngster that immediately attacked him and inflicted a wound in his back which was slow to heal. The male defended himself as best he could and retreated to his hutch at the first opportunity. The mother did not participate in the fracas. From that time she became more tolerant of the male, and by early spring he was accepted as a member of the group, all six often playing and tumbling in the water together.

Most imposing of the otters is the giant otter of the great river systems of eastern tropical South America, with reports of old males 7 feet, 2 inches long. Our first specimen came in 1955. This was a well-grown but immature female, weighing a little more than 28 pounds and measuring 43 inches in total length. She was perfectly tame and gentle and obviously accustomed to human companionship. She was kept indoors in a large exhibition cage with a glass-fronted pool and an elevated duckboard on which she could dry her coat. She spent much of her time in the pool, her great agility in making twists and turns giving the impression that she had plenty of exercise space. By the time she had grown to 56 inches and had increased considerably in bulk, however, we felt that she really should have larger quarters. A large outside inclosure equipped with pool and shelter was built and with some difficulty she was moved into it. All our efforts were in vain, however, for she protested continuously in a loud, whining voice, refused her food, and paced to and fro, without ceasing, along the restraining wire netting. After a few days we gave in and returned her to her accustomed, if cramped, quarters, to her obvious satisfaction.

The aristocrat of the otter family is the sea otter, once hunted to the verge of extinction for its exquisite fur but now slowly increasing in numbers under strict protection. One race extends from Kamchatka to the western Aleutian Islands and south to the coast of British Columbia, while another is found locally from the coasts of Washington to Baja California. Most of its life is spent in the water, although during stormy weather it may come ashore, where it is awkward and slow-moving. In size, it compares quite well with the giant otter.

Many naturalists have described the sea otter's habit of floating on its back and of breaking the hard shells of the shellfish on which it largely feeds by pounding them against stones held on its chest.

The northern sea otter is said to feed principally on sea urchins, with mollusks, crabs, and the like making up approximately a quarter of its diet, crabs about 10 per cent, and fish even less. The southern race apparently prefers abalones. None of these items are particularly easy for most zoological parks to obtain, and thus difficulty might be expected in keeping the animal in captivity. After a few false starts, however, when much was learned about the temperature tolerances of the animals—which are not very great—it was found that various kinds of fish, land crabs, clams, and frozen squids are acceptable food. Sea otters may never become common zoological exhibits, but it is a hopeful sign that the Wood-land Park Zoological Gardens in Seattle has had outstanding success with them and kept one elderly female, known as Suzie, for 6 years, 17 days.

THE HYENAS

A hyena is not an animal that would seem likely to inspire affection in human beings. In all three species—the spotted or laughing hyena of Africa, the striped hyena of Asia and Africa, and the brown hyena of South Africa—the foreparts are heavy and powerful, the shoulders sloping sharply downward to the short and relatively weaker hind legs. The jaws and teeth are immensely strong, capable of crushing even heavy bones with ease. Nocturnal, skulking, and "cowardly" in habit, the hyenas are scavengers, relying for their food largely on the kills of bolder carnivores, although they are not averse to living prey too weak to make strong resistance and frequently devour young or injured game or domestic stock. Even sleeping men are not immune from attack, and there are records of children having been carried off by hyenas.

Nevertheless, despite their unsavory reputation, hand-reared hyenas under favorable conditions may remain perfectly tame, even after they have become adult. There is on record an account of a female striped hyena that, after 13 years, still remained a household pet.

We have had an embarrassing experience with such a pet animal in the New York Zoological Park. It concerned a young spotted hyena that we purchased from the San Diego Zoological Garden, where it had been hand-reared and exhibited as a pet in the children's zoo. While we were quite aware of the animal's gentleness and friendliness, in discussing the new arrival with a newspaper reporter our curator of mammals talked at some length about the habits of hyenas in the wild—quite naturally using such terms as "skulking" and "scavenger." The reporter—also quite naturally—quoted the curator along these lines. An abbreviated version of the story was sent out by a wire service and the following day was published in a San Diego newspaper. This condensed form made it appear that our curator was calling the beloved children's zoo pet a "skulking scavenger," and there was an immediate outcry from San Diego children. According to a story wired to New York newspapers, the children were contributing money to a fund for repurchase of the animal and demanding its return to San Diego, since the New York zoo did not love their pet! Our curator hastily issued a statement, saying that this was the nicest, friendliest, gentlest, most intelligent hyena he had ever seen, that the New York zoo considered itself lucky to have such a delightful animal, and that of course *this* hyena was not a "skulking scavenger." San Diego was thereupon appeased.

As a matter of fact, our hand-reared specimen does appear to be perfectly tame, gentle, and friendly, at least with its keepers and members of the mammal department staff, and when one of them appears, it is likely to come to the front of its cage, pressing against the wire and waiting to be scratched, its lips drawn back in a characteristic "smiling" position. On rare occasions, especially when food is momentarily offered and withdrawn, it will utter the chattering cry that has often been likened to a laugh.

A curious misunderstanding, apparently of ancient origin, concerning sex in the spotted hyena still persists. This is the belief that an individual animal may change from male to female and vice versa. The confusion has arisen from the fact that the external sex organs of the virgin female are superficially identical with those of the male and show only slight changes after the birth of young, beyond enlargement of the nipples. The peculiarity is confined to the spotted hyena, and the sexual differences in striped and brown hyenas are easily recognized.

Since they are naturally shy and retiring, as well as essentially nocturnal, hyenas are inclined to keep out of sight during the day, and if forced to expose themselves to light, they may become restless and overexcited. There are, conse-

Hyenas are primarily scavengers and are often called "cowardly" because they usually feed on the kills of larger predators. This is a spotted hyena.

quently, many advantages in exhibiting a hand-reared specimen, such as our San Diego example, which is quite accustomed to people and is active most of the day. The hardiness of hyenas is somewhat surprising, considering their tropical habitat. While ordinarily hyenas are kept in heated quarters in the colder areas, a brown hyena of unknown origin once lived out-of-doors in the New York Zoological Park for nearly 12 years in a run approximately 25 by 50 feet which was equipped only with a small, unheated shelter. Temperatures dropped as low as −14° F. during one winter, with no apparent ill effect on the animal.

Hyenas have an excellent longevity potential in captivity and there is a record of a spotted hyena that lived in England for about 25 years. Several striped hyenas have lived in captivity for more than 20 years, but the best record for the brown is 13 years, 5 months, 10 days, in the Zoological Gardens of London.

THE CATS

The cats are among the mainstays of zoological gardens. Widely distributed over most of the world, except the Australian region and Madagascar, they are most numerous in warm areas. In size they vary from the lion and the tiger to the tiny black-footed cat of South Africa, which is no larger than a small domestic cat. Nocturnal habits are the general rule in the family, although occasional individuals of most of the species may prowl by day. Many kinds of cats are at home in the trees, but the lion, the tiger, and the cheetah are essentially terrestrial. There are curious variations in attitude toward water. The lion, the leopard, and many of the smaller species are reluctant to enter it, although all, of course, can swim if they have to. On the other hand, the tiger and the jaguar, at least, will go into water freely and swim with apparent pleasure. At one point when we were exhibiting hand-reared tiger cubs, their attendant routinely entered their cage and played a hose on them on hot summer days. The cubs' delight in playing with the stream of water was obvious.

Handling of the cats, especially the larger ones, used to be a greater problem in zoos than it is today. Many types of "squeeze cage" have been evolved for controlling the animals for minor operations, usually on teeth or claws. When our present lion house was opened in 1903, it was provided with a traveling shift cage, by means of which animals could readily be moved from one cage to another. One side of this cage could be moved toward the other, thereby reducing the space so that the inmate was eventually immobilized. Even in the early days "squeezing" was rarely undertaken, largely because of the danger of self-inflicted injury by the frightened animal. Today even the largest cats can be so readily immobilized by drugs now in general use that the "squeeze cage" no longer seems to serve a useful purpose.

The great scourge of cats in captivity is "cat enteritis," or feline panleucopenia. The ever-present domestic cat provides the great reservoir for the transmission of this usually fatal disease, so that, at least in this area, it is impossible to avoid occasional exposure. For this reason all felines, young or old, immediately after arrival, are started on a course of treatment with a commercial vaccine.

As is well known to breeders of domestic cats and to most zoo people, cats are especially sensitive to coal-tar products that may cause fatal results if they come into contact with the skin. Cresylic acid seems to be the responsible agent.

According to one widely accepted grouping, the lion, the tiger, the leopards, and the jaguar are classified as the roaring cats; all the smaller cats from the puma down, with the exception of the cheetah, are the purring cats, and the cheetah is in a group of its own. The best-known member of the purring cats is, of course, the domestic cat. Descendants of local races of the kaffir cat and the jungle cat, domesticated and held sacred by the ancient Egyptians and carried, in the course of time, to Europe and Asia, are believed to have been interbred there with indigenous species, so that our familiar household pet presumably has a heterogeneous origin.

Closely related to the domestic cat and suggestively similar in color to the striped tabby pattern of domestic cats, but with shorter and thicker tail, is the European wildcat, which still exists in Scotland and extends across Europe to Asia Minor. It is reputed to be fierce and untameable, but it certainly could be no more savage than a truly feral domestic cat. The only specimen we ever exhibited was noticeably quiet, but that was perhaps because special efforts were made not to disturb it.

The tiny leopard cat, which has an enormous range from eastern Siberia west to Baluchistan and south through India, the Malay Peninsula, and the larger islands of the Philippines, has a reputation for untameable ferocity, but hand-reared specimens might be more responsive. A female kitten of this species, no more than a few months old, was received some years ago. She was extremely savage and could only be kept quiet by providing a darkened sleeping box in which she spent most of the day. During her quarantine, while receiving her anti-feline panleucopenia treatment, she went into convulsions so frequently that the chances of reconciling her seemed slim. When she was finally removed to a glass-fronted cage in our small mammal house, however, she quickly learned to enter a shift cage when the door was opened by remote control, while her quarters received their daily cleaning. She soon became so accustomed to this routine that she spent her days sleeping quietly on a large log and had no further occasion to exhibit savagery. Another indication of the individual variation in animals was that while most small cats seem to require a diet consisting largely of whole small mammals and birds, with very little beef or horseflesh, this leopard cat ate only four to eight ounces of whole raw horsemeat daily and consistently refused rats, mice, rabbits, and pigeons.

We have never exhibited the smallest of the cats, the tiny black-footed cat of South Africa, whose length, including the long tail, is only about twenty inches. It is seldom imported.

By contrast, the puma—known also as the mountain lion, cougar, panther, and catamount—is probably more often seen in zoological gardens than any other of the large cats except the African lion—a situation which by no means reflects its status in the wild. This great, solidly brown, tawny, or grayish cat once ranged over most of North America and South America from southern Canada to Patagonia. In North America it is now restricted, except for possible stragglers, to Louisiana and Florida and the wilder parts of the west, from southern Canada to Mexico. A puma killed in Maine in 1938 is believed to be the last taken in the northeast.

As in most members of the family, males exceed females in size and may weigh as much as two hundred pounds and occasionally even more—two hundred and seventy-six pounds is reported for an Arizona puma. A cat of such size is, of course, an animal to be reckoned with and if there actually are stragglers left in the wilder parts of the northeastern states —as rather inconclusive newspaper stories sometimes intimate—they might well be feared or at least carefully avoided! Deer seem to be the puma's favorite food, although domestic stock may pay a heavy toll in some parts of its range. It seldom becomes a confirmed maneater, although there are

The ocelot, found from the southern border of the United States to northern Argentina, is a tame and attractive pet when young, but is likely to be unpredictable and even dangerous as it matures.

authentic accounts of persons having been attacked and even eaten.

Pumas take readily to captivity conditions. Those from the cooler parts of their range, at least, are indifferent to the cold. Most zoo specimens have been bred in captivity or reared by hand from an early age, so that they are fully accustomed to restraint, and food is of course no problem. A female Rocky Mountain puma, largest of the North American pumas, lived with us more than ten years on a diet of four pounds of whole raw horsemeat into which fine bone meal and cod-liver oil had been scored; she received this diet six days a week, being fasted on the seventh day, as are all of our larger cats.

Puma kittens are delightful little creatures. There may be any number from one to four, and they are spotted at birth, with ringed tails.

Every zoological garden, I suspect, can match our experience of telephone calls and letters inquiring about the suitability of the smaller New World cats as pets. The animals referred to are usually the ocelot, which is distributed in forested areas from the southwestern borders of the United States to northern Argentina, and the closely related and dainty margay, which ranges from southern Texas to northern Uruguay. Both animals are frequently taken when very young and reared by hand, so that they make tame and attractive pets. As they approach maturity, however, they are likely to become unpredictable, even dangerous, and serious accidents have occurred when such animals have been privately kept. To all inquirers we have a standard reply in which we strongly advise against such wild pets.

Interesting as the smaller cats are, to the zoological garden it is the great cats—leopard, lion, tiger, jaguar, clouded leopard, and snow leopard—that are of the greatest importance. Traditionally these animals have been housed in large buildings, variously known as the lion house, the great cats house, the carnivora house, and so on. As befits what is inevitably a focal point in any zoo, the buildings are usually architecturally imposing. The basic plan is generally a walk-through with barred cages on one or both sides, connected with outdoor compartments for use in clement weather. A typical example is the lion house in the New York Zoological Park, opened to the public in 1903, then an advanced model of its kind and still an excellent, serviceable building. It is two hundred and forty feet long and one hundred and ten feet wide, with six interior cages measuring eighteen by twenty-two feet and six smaller ones measuring twelve by twenty-two feet. These connect directly with an outdoor series, the largest of which are thirty-eight by forty-two feet, six inches. The larger outdoor cages are backed with rockwork and have small but deep pools for the use of species that enjoy bathing. All cages, both indoors and out, are supplied with the scratching logs essential for keeping claws in order.

The most important innovation in 1903 was the use of wire mesh instead of bars for cage fronts. This was installed in spite of wide objection and, it must be said, with some qualms, on the basis that wire would never restrain lions and tigers. The wire used was five-gauge steel, American Standard, and the mesh was three inches. It is still in use and slight replacements have been required only because of rust damage. It has never been damaged directly by any cat, nor has it ever permitted any animal to escape. It is my personal feeling that this wire, painted black, provides better visibility than bars can possibly give.

Nowadays the new forms of glass, such as the almost unbreakable Herculite, permit a different type of cage front, but glass has size limitations and for very large areas the old steel mesh insures complete safety without the heavy appearance and unpleasant implications of bars.

Originally our indoor cages were floored with oak two by four strips, set on edge. These eventually became so saturated with urine, however, that they had to be replaced—the beginning of a long series of trial-and-error experiments. When wood was abandoned, the floors were covered with very fine steel-troweled concrete. In use for more than 30 years, this material has caused no trouble by abrading the feet of the animals. An oak duckboard, measuring four by six feet, is placed on the floor of each cage to insure a warm and dry sleeping place. This reversion to concrete is, of course, in direct opposition to the long-established feeling that floors for big-cat cages must be of wood. Wood quickly becomes foul-smelling, however, an objection that outweighs any favorable qualities it may have. The new lion house in the Jardin des Plantes in Paris has floors made of grooved teak blocks, possibly an effective but presumably expensive solution.

Unless the ancient custom of poking meat through the bars with a long-handled fork is to be followed, a better and safer method for feeding the big cats must be devised. In our lion house there is a space about five inches high beneath the wire at one end. This opening is closed by an iron bar that is removed at feeding time, so that the keeper is able to push the meat through without risk. This method, clumsy as it may be, has served us well. The only improvement I have seen is a device in the Jardin des Plantes. This consists of a V-shaped steel trough, hinged at the bottom, which can be tipped toward the keeper's space for filling, then inward to the cage, one side always closing the opening. The trough can be locked, of course, in either position. This tipping trough is identical in principle to the in-and-out aperture at the "drive-in" window of some banks!

The open-moat method of exhibit, based largely on the experiments of Carl Hagenbeck at the beginning of this century, is now widely used. The moat is effective for both lions and tigers, but is less practical for the smaller members of the cat family because of their leaping and climbing ability. I have not seen a successful open inclosure for these animals.

The salient point, of course, is the width and depth of the surrounding moats. Hagenbeck suspended a stuffed pigeon ten feet above the ground and allowed various cats to attempt to bring it down. He found that lions and tigers could jump upward only about six feet, six inches, and that leopards could not quite reach the pigeon. These figures were once confirmed to me by the late Theodore Schroeder, a former Hagenbeck employee. Nevertheless, they seem to me somewhat short of the animals' leaping ability, as I have seen a tigress, carrying a ten-pound piece of meat, leap cleanly to a shelf eight feet above the floor, and have measured the claw marks of a leaping lion, left twelve feet, four inches above the ground level. Hagenbeck also found that both tigers and leopards could cover ten feet, on the flat, from a standing start, and he felt that both might have done thirteen or fourteen feet with a run. He therefore made his trench at Stellingen twenty-eight feet wide, a figure somewhat reduced in later practice.

Noted examples of the moated inclosure for lions and tigers in America are those in the Detroit and Chicago Zoological Parks. In the former the water moat used for confining tigers is twenty-five feet across and sixteen feet deep, while that for lions is twenty-one feet wide by twelve feet deep. In Chicago the dimensions for tigers are the same as those in Detroit, but the Chicago lion moat is twenty feet across and sixteen feet deep. Detroit made the slightly concave back wall of both lion and tiger exhibits twenty feet high, while

Young male lions for a long time lived peacefully together on Lion Island in the New York Zoological Park.

Chicago's lion wall is only sixteen feet high. In use for enough years to establish their safety, these dimensions are now generally accepted as standard.

Our own lion rock, opened in 1941, comprises an exhibit area roughly triangular in shape and measuring about fifty feet on the viewers' side and approximately sixty-five feet to the apex. At front and back it is inclosed by dry moats twenty feet wide and sixteen feet deep, while at the sides wooden palings rise to a height of 16 feet. At the left, these palings conceal a low building containing seven individual cages to which the animals are confined during periods of snow and cold. Since there is no backstop and the front moat is concealed by low shrubbery, the lions appear to be entirely free. Beyond them, antelopes may be seen grazing quietly on the African Plains, and from the opposite direction, with the antelopes in the foreground, the distant lions appear to be integrated in the same exhibit. As first constructed, no provision was made for the return of an animal that fell or jumped into the moat. When such an accident finally happened, the unfortunate lion could not jump or climb back to the top of the exhibit rock and was securely trapped. As an emergency expedient, a wooden ramp was quickly built and lowered into the moat at a concealed point, giving access to the shelter building. This contrivance was barely in place before the frustrated lion climbed it, and the ramp has now become a permanent part of the operation. It was, in fact, both used and misused, several times during the first year of operation of Lion Island. One lion in particular seemed to land in the moat frequently and the first time or two, although he climbed the ramp readily enough, he was reluctant to jump through the door at the top of the ramp so as to come out on the island. A piece of meat suspended just outside the door was enough to make him decide to make the final jump. After several repetitions of this behavior, it appeared to the keepers that the lion was purposely jumping down into the moat, running up the ramp, and waiting at the door for meat to be offered. The next time he did so, a concealed keeper reached down from behind the lion and swatted him hard on the tail with a broom. The lion leaped forward, the door slammed behind him, and the game was over.

The most recent development in exhibition of the big cats brings the formal building and the moated outdoor area into a single unit. This has been accomplished in the Woodland Park Zoo in Seattle and in the Philadelphia Zoological Garden. Each of these fine buildings has two open-moated areas, one for lions and one for tigers, to which the animals may be admitted from the indoor exhibition cages. Another innovation, perhaps of greater importance, is the use of Herculite for cage fronts. This glass is built in three layers, with an over-all thickness of three-fourths inch and a calculated resistance to a blow of three hundred pounds delivered at a speed of thirty miles an hour. As a safeguard in the unlikely possibility of complete collapse of the glass, an electrical trip installed in the middle layer of plastic will drop a steel-mesh panel across the opening.

The leopards of Asia and Africa present the interesting case of an animal with "normal" ground color of tawny-yellow, marked with rounded black or dark brown spots, arranged in rosettes, and at the same time a melanistic, or black, tendency so that black individuals are found with some frequency in parts of Asia; black leopards are rare, but not unknown, in Africa. Both normally colored and melanistic examples are said to occur in the same litter in nature. In my own experience, captive pairs in which both animals were black have produced only black young, and I do not recall an instance in which black young have been born to spotted parents. A pair of leopards in the Zoological Gardens of Copenhagen, one spotted and the other black, bred cubs of both color phases, however.

The leopard is a bold and resourceful hunter, lurking in the foliage of forest trees or springing from ambush on the ground. It feeds on animals up to the size of deer, antelopes, and domestic cattle. Although it is not generally given to attacks on man, there are numerous accounts of maneaters —one, indeed, of an Indian leopard credited with killing more than two hundred persons within a three-year period.

Our most revealing experiences with black leopards began with the arrival, in the spring of 1941, of a pair of adults that had become well adjusted to each other. The male was a rather quiet, timid animal and the female, bold and aggressive in the true tradition of the supposedly more dangerous black leopard, was completely dominant. A single cub was born to this pair in 1942 and a litter of two in 1943. During the entire period the parents continued to live together. When birth was impending, the female retired to the den, which she did not permit the male to enter. When the cubs finally wobbled into the outer cage, she continued to protect them, but as they grew stronger the male was gradually allowed to become familiar with them and finally to indulge in strenuous play. During early cubhood the male was never seen to make an offensive gesture toward the cubs, although this may have been due to the defensive attitude of the dominant mother. When the youngsters reached the age of approximately nine months, the male began to show impatience of their lively antics, and when he was finally seen to cuff them severely, it became necessary to remove them. This was a serious step, for the cubs, suddenly deprived of the protection of their parents, were as frightened as any wild-caught animals, and became reconciled only after a considerable time.

Because of this difficulty, it was decided that the next litter should be hand-reared. Therefore when a single cub was born in 1945, it was promptly removed and turned over to Mrs. Helen Martini, wife of the lion house keeper. This cub, like its brothers and sisters, was black and naturally was named Bagheera. At birth he was fourteen inches from tip to tip and weighed twenty ounces.

The first weeks of Bagheera's life were spent in the Martini home across the street from the zoological park and on some evenings, after the close of the zoo, Mrs. Martini used to take him on a stroll through the grounds, on a leash. His temperamental progress was most interesting. As a small cub, his disposition was friendly enough, as might have been expected, but gradually he began to concentrate on the Martinis and to withdraw from others. When he had reached his maximum recorded weight of ninety-eight pounds, at the age of ninety-six weeks—by which time further weighing was thought to be no longer safe or practical—Bagheera was removed to a cage in the lion house and the Martinis no longer entered it with him. Completely friendly relations were maintained through the wires, however. Other keepers or workmen who crossed the guardrail were cautioned to give Bagheera's cage a wide berth, for a black paw would shoot through the wire like lightning if there seemed to be the slightest hope of striking home. Both the Martinis, on the

A black leopard cub, named "Bagheera," was hand-reared by Keeper Helen Martini in her home.

other hand, could stand beside the cage and reach through and pet the animal.

When the big cats are mentioned, the lion is probably the one that more often comes to mind. "King of the beasts."

And yet the very anthropomorphizing that long ago bestowed this title on the lion has been employed, in later years, to dethrone him. A "king" has certain standards from which he may not deviate, so that when a starving lion soothes his hunger by eating small rodents or even carrion his position comes into question. But the lion, despite arguments that are no concern of his, is still an impressive animal. His most striking characteristic, I think, is that he is unafraid. A bold, powerful hunter, with no enemies of consequence, he has no sly tricks but goes straight to his business of firm attack. Zebras have usually been rated as the African lion's favorite prey, but a check made in the Kruger National Park in the Transvaal revealed that wildebeests were first, waterbuck second, and zebras third. Five buffalo, two giraffes, a young hippopotamus, and an ostrich are also listed as victims. Since food animals of these types are mostly dwellers on open plains where grass is abundant, the lion lives there also, avoiding large deserts and dense forests.

This way of life brings the lion into contact with man and his herds of domestic stock that graze over the same lands as antelopes and zebras. Depredations are a natural result, and man himself has not been exempt from the toll. Stories of the misdeeds of maneating lions have produced some exciting books, but it is still something of a shock to realize that in these days the loss of human lives to wild animals is almost a commonplace. But there is no doubt about it. The Game and Fisheries Department of the Uganda Protectorate reported at least seventeen deaths of natives from attacks of lions in the period from January, 1955, to June, 1956.

The lion once ranged across the plains of western and central India, southward from Sind, and from Greece, Turkey, and Persia to Africa, where it avoided only the Sahara and the great forests of the west. Today it probably does not exist outside Africa, except for the remnant of the Indian lion under strict governmental protection in the Gir Forest of Kathiawar in India—a matter of perhaps two to three hundred animals. In Africa itself the lion has been eliminated from the extreme north and south and elsewhere its numbers are becoming reduced, for as the game goes, so must the lion.

The physical characters of lions vary greatly. The mane of the male makes its first appearance at the age of about eighteen months, as a ridge along the nape, growing gradually downward and backward, reaching full development at five to six years of age. It may be tawny, dark brown, or black, and may be very full with extensions along the belly line—or it may be scanty or even lacking entirely, for maneless males are not uncommon. The "thorn" concealed in the tuft of hair at the tip of the lion's tail is a horny attachment of the skin with no known function.

Living male lions weighed in the New York Zoological Park have ranged from 330 to 410 pounds, and a single female was 250 pounds. Even heavier weights have been recorded in the wild—14 freshly killed males weighed from 329 to 421 pounds, and five females from 269 to 409 pounds. The greatest weight for a male is believed to be 516 pounds.

Zoo lions have been bred in captivity for so many generations that physical changes are likely to occur. Shortness of leg, "pushed-in" face, known as "bulldog," and sway backs are some of the signs that denote degeneration, but if breeding stock is chosen carefully really magnificent animals can be produced. Breeding in captivity is commonplace today, but it was once an undertaking of consequence, in which the Zoological Gardens of Dublin and Leipzig, especially, achieved fame. Between 1857 and 1950, Dublin produced 502 cubs and in a much shorter period, from 1878 to 1912, Leipzig had about 700 lion births. In their heyday these great centers supplied lions to the zoological gardens of the world, but gradually every zoo of any size began to produce its own cubs and nowadays few attempt to breed lions, except on rare occasions, because of the problem of disposing of the weaned cubs.

Moated areas are an excellent ways of exhibiting lions, but thought has to be given to the animals that must live together. They must get on well together, and this means that males are the more useful and that no females can be included, as otherwise serious fighting will result. The best method is to secure the required number of sound, well-reared cubs, aged from 6 to 9 months, and let them grow up together. Even then there may be problems. On our lion island, all went well with a "pride" of five males until they reached sexual maturity, when an unusual amount of quarreling and segregation occurred. We found that a rather small, pale-colored animal had developed a mate-relationship with the dominant male and was kept apart from the group by his protector, which would not allow another male to approach. This situation finally settled into a routine, accepted by all, with a minimum of disturbance. It continued until the sudden death, some two or three years later, of the dominant male. After that time the surviving member of the couple was accepted as a member of the group on equal terms with the others and did not make another attachment.

When these animals were 10 years old, we considered them to be too inactive to make a good show, and since a group of male cubs was not available at the moment, we acquired a young pair. At the age of about four years, the female produced two cubs. We had some doubts about allowing the cubs the freedom of the island, for fear they might fall into the moat. When they were liberated with their parents in the early spring, however, the solicitude of their mother removed all our fears, for she would not allow them to approach the brink. Either by nudging them away from it or interposing her body between them and the edge, she guarded them from a danger she obviously recognized. In time, as they grew older, they were given more freedom by the mother, but by then they had been conditioned to avoid the brink.

It is too bad, in a way, that disposal makes it so impractical for zoos to produce a continuous sequence of cubs, for the babies are undeniably appealing. At birth they are more or less strongly spotted and striped, the dark markings sometimes persisting into adulthood. The statement has often been made that lion cubs are born with their eyes fully open. This may happen on occasion, but certainly all of the many newborn cubs I have seen had their eyes closed. I recall two female cubs, both of which opened one eye on the fifth day and the other on the next.

Each of these cubs measured 21½ inches, tip to tip; one weighed 2 pounds, 12 ounces, the other an even 3 pounds. A male born dead in the same litter weighed 3 pounds, 15 ounces.

From records kept in zoos, we know a good deal about the life span of lions. Somewhere between 10 to 15 years, signs of senility appear, but there are plenty of records well over 15 years. One of 29 years in the Cologne Zoological Garden apparently lacks full documentation, but the 25 years, 18 days achieved in the Dublin collection is certainly authentic.

THE TIGER

Up to 1944 our experience in breeding tigers had been anything but good. From time to time cubs had been born, but no mother had ever succeeded in rearing her offspring. In that year we had a small, very nervous, jungle-bred female named Jenny that had been with us for 10 years, paired with a very large but decrepit male. Jenny had produced several litters of cubs but had lost them all. We determined, therefore, to lift the next litter and attempt hand-rearing.

And so when, on February 8, 1944, small squalls were heard coming from the cubbing den in the lion house, we coaxed Jenny out of the den and entered it from the back. Wearing cotton gloves well soaked in Jenny's urine and carefully dried, Keeper Fred Martini picked up three beautiful cubs from the nest of straw in Jenny's den, and called out their sexes, weights, and measurements. Soaking the gloves in tiger urine was, of course, a precaution against contaminating the cubs with the human smell.

Proceeding cautiously, we decided to remove only one cub at this first view and to see whether Jenny would do better as a mother than she had previously. But in the next two days the sounds from the den became weaker and weaker, and it was obvious that Jenny had not reformed, so we lifted the remaining two cubs and turned them over to Mrs. Helen Martini, the wife of the lion house keeper.

Mrs. Martini's home was across the street from the zoo and the cubs were taken there, where preparations for just such an eventuality had been worked out well in advance. At first they were kept in a box about 2 feet in each dimension, with open top and padded sides and floors. An electric heating pad was attached upright on one side to provide warmth without risk of overheating. It was soon found that the cubs had a tendency to attach their mouths to the feet or tail tips of each other, sucking vigorously, so that separate boxes had to be provided.

Caring for the cubs was a full-time job for Mrs. Martini, as their baby formula of evaporated milk, water, multiple vitamin preparation, and dicalcium phosphate had to be given them at 3-hour intervals from 6:00 A.M. to midnight.

The cubs were eight weeks old when we undertook to experiment with a more substantial diet. Each cubs was given a small ball of chopped raw meat. They ate it greedily, but the results were appalling: severe gastric disturbances—in one case so severe we feared the cub might not recover. They soon returned to normal, however, and we substituted meat juice for the liver extract that they had been getting for the three weeks previous to our ill-timed experiment. We waited until they were just short of 11 weeks old and tried again, this time offering a section of rib to which bits of raw meat adhered. The cubs eagerly licked these "as clean as a bone," and since they showed no ill effects, we gradually increased the quantity of raw meat. It was not until they were 18 weeks old that they could be induced to drink by lapping, and even then they did so reluctantly.

At their birth, each of the three cubs measured exactly 20 inches from tip to tip, and their eyes were closed. Mrs. Martini was given the pleasant chore of naming the babies and from an atlas of India she chose the names of three towns. The female, which weighed 2 pounds, 11 ounces at birth, she named Dacca; the largest male, 2 pounds, 9 ounces, was

"Rajpur" was one of three tiger cubs hand-reared by Keeper Helen Martini. Until the cubs were nearly half-grown, Mrs. Martini played with them and hand-fed them.

named Rajpur; and the slightly smaller second male, 2 pounds, 8 ounces, she named Raniganj. Dacca opened one eye on the ninth day, Rajpur on the eleventh, and Raniganj not until the seventeenth day. Usually the second eye opens the day after the first.

Even in the wild, tigers have strong temperamental differences and local villagers in India are said to know "that such a one is daring and rash; another is cunning and not to be taken by any artifice; that one is savage and morose; another is mild and harmless." We noticed very obvious characteristics in our cubs at an early age. Recording their development in the New York Zoological Society's magazine, I noted at the age of four to five weeks Raniganj was "definitely bad tempered," Rajpur was "fat and indolent," and Dacca was "bright and friendly." Four months later I referred to them again: Raniganj was "furtive and unpleasant," Rajpur was "slow, fat, lazy, and good-natured," and Dacca "remains the hoyden." It is interesting to observe that these characteristics, in more mature form, of course, persisted when they became adult.

When the cubs were about 3 years old, Dacca began to show a strong preference for Rajpur and made life so unpleasant for Raniganj that he was removed to another cage. Dacca reached menarche at about 3 years, 8 months, and sought the attentions of Rajpur, rolling on her back, licking him, and striking him playfully with her paw. Copulation was not observed, however, and so we were taken by surprise one May morning in 1948 when she gave birth to a cub in an open cage. As we watched, she leaped to a shelf and dropped another that died later of a ruptured spleen. Dacca was quickly moved to the adjoining cage and the cubs were placed in a nest box in the den. She accepted the change readily and later in the day gave birth to two more cubs.

From the first Dacca showed the greatest devotion to her cubs and gave no evidence of disturbance at being fastened out of the den each day while Mrs. Martini checked and handled the cubs. In fact, on one occasion when Mrs. Martini was calling to her from the cage front, Dacca brought a cub in her mouth and placed it in Mrs. Martini's hands extended into the cage through the wire front. Dacca then walked toward the den, but when Mrs. Martini called frantically for her to come back and take the cub, she calmly did so and returned it to the den.

Between 1948 and 1959 Dacca produced 11 litters (she was not bred in 1955) totaling 32 cubs, of which 28 were fully reared, some by Dacca and some by Mrs. Martini. The litters varied from one to four and there were 19 males and 13 females. We were interested to see that despite the fact that this was a brother-sister mating, all the cubs were strong and vigorous.

Young tigers are not so difficult to dispose of to other zoos as young lions, and we had little trouble in placing Dacca's offspring with zoos in this country, Europe, and Australia, although by the time the thirty-second one came along we felt that we had just about saturated the market and perhaps it was time to stop. Dacca apparently felt about the same way, for while she was still a good mother, she was noticeably less attentive to her final cub, which, incidentally, we named Finis.

Dacca, "the hoyden, always sweet," kept her gentleness and friendship with the Martinis and other members of the New York Zoological Park staff to the very end. In September, 1964, when she was 20 years, 7 months, and 2 days old and obviously failing fast, she was painlessly put to sleep. Her life span was the longest on record for a tiger up to that time. Similar euthanasia had been decreed for her mate, Rajpur, in 1961 when his health failed. Previously Raniganj had been disposed of to an American dealer in animals. Of all her cubs, the zoo retained only one, a daughter, Dacca II, of her sixth litter, born in 1953.

Occasionally a zoological garden exhibits one of those unnatural curiosities of captivity, a "tigon" or a "liger"—the former being a hybrid between a male tiger and a female lion ("tiglon" is another name), the latter between a male lion and a female tiger. Such crosses are improbable in nature, since there are no tigers in Africa and lions in India are confined to a very small area, but they are not uncommon with captive animals. The Hagenbecks bred many of them in Germany in the early years of the century, including one male hybrid that weighed as much as both its parents together, an increase in size apparently being characteristic of such animals. Young ligers are strongly spotted and striped and some of these markings usually persist in the adult. Incidentally, while these hybrids are usually sterile, the females at least are sometimes fertile.

THE JAGUAR

The jaguar is not only the largest member of the cat family found in the Americas, but worldwide is exceeded in size only by the lion and the tiger. It ranges from the southwestern borders of the United States through Mexico and Central America to northern Patagonia and its size increases toward the south. At the northern extremity of its range it is not a particularly large animal, probably seldom reaching a weight of 250 pounds, but Sasha Siemel, whose jaguar-hunting exploits are well known, once told me that he had killed male jaguars in the southern Mato Grosso that weighed more than 350 pounds.

In captivity the jaguar seems more inclined to be morose, even occasionally savage, than does the more alert and active leopard. This tendency persists even in hand-reared animals.

Well-established pairs of jaguars will breed regularly in captivity, and more than one small zoological garden has financed its animal purchases through the sale of the offspring of such a pair. In establishing reliable breeders, the best results are certainly obtained by introducing young animals and allowing them to mature together, for strange adults are not always amenable. We found this out to our cost in the early days of the Zoological Park when we admitted a beautiful young female jaguar to the cage of an apparently friendly male. For some days the two had been able to see and smell each other through a barred interconnecting door and gave every evidence of wanting to carry their acquaintance further, so that we were quite unprepared when, the instant the female walked into the cage of Señor Lopez, the male, he sprang on her and with one bite killed her by crushing two cervical vertebrae.

Nearly fifty years passed before we at last acquired a breeding pair of jaguars, litter mates born and reared by hand in the Cleveland Zoological Garden. When they were a little more than three years old, a single male cub was found in the den one morning and the father was promptly removed. The cub appeared to be normal, 16 inches long, weighing 1

Keeper Helen Martini hand-reared these jaguar cubs whose own mother was nervous and unstable.

pound, 14½ ounces. Its eyes opened three days later and all seemed to be going well, but on the following day it could not be found, presumably having been devoured by the mother.

For two weeks following the birth, the jaguars remained separated and during this short time the male developed marked truculence, even toward his keepers. When we finally allowed him to reenter the female's cage, he was still so aggressive that he could be left with her only a few minutes. Things got better within a week or two, however, and several matings were observed, and eventually, after a gestation period that was something between 98 and 109 days, one male and two female cubs were born. Again all seemed to go well for a few days but after a week the mother began carrying the cubs around the cage and showing signs of uneasiness. This time we knew what might happen and immediately removed the cubs and gave them to Mrs. Martini to hand-rear.

THE SNOW LEOPARD

Certainly among the finest of the great cats is the snow leopard or ounce. "Snow leopard" is an appropriate name, for it comes from heights of 6,000 to 18,000 feet in central Asia, from Tibet and Turkestan north to the Altai Mountains, climbing far above timberline in summer and staying in the snow belt throughout the year.

The snow leopard has always been uncommon in captivity, because of the remoteness of its habitat and its sensitivity to extreme heat. Nowadays, when animals can be shipped quickly by air at almost any season of the year, only remoteness is much of a barrier.

A favorite snow leopard in the New York Zoological Park was a lovely, tame male that was acquired by us not long after the end of World War II. Bowser, as the animal was named by his owner, had been acquired as a kitten by an American Air Force pilot and had accompanied him on flights "over the hump" in Burma. Brought to this country by his owner after his discharge, Bowser soon became "too much animal" and ended up in the zoo. He remained quite gentle during his stay in the zoo and always welcomed visitors to his cage by purring audibly. The moment a visitor started to leave, however, Bowser's temper changed. He would then wind himself around the visitor's feet, biting and clawing with mounting vigor. For this reason all petting and stroking soon took place only through intervening bars.

Bowser was kept, the year around, in one of the outside cages of our lion house. During cold weather he always appeared content, but in the high temperatures of his first summer he gave indications of distress, breathing heavily and eating little. A mere change to another den, cooled by air from below, brought him back to a more normal condition.

Bowser lived with us 8 years, 8 months, 15 days, apparently a record by about two months. Unfortunately, we had no female during this period, for such a gentle and friendly— when not overly friendly—animal might have made an excel-

"Bowser," a tame and gentle snow leopard, originally belonged to an Air Force pilot and accompanied him on flights "over the hump" in Burma.

lent parent. There are several records of the breeding of the animal in captivity, but rearing is comparatively rare.

THE CHEETAH

The cheetah has been called "the mildest cat," but I can recall at least one that was anything but mild. A young female, perfectly gentle toward her keeper, for weeks would have nothing to do with an equally young and, we would have thought, attractive male cheetah to which we tried to introduce her. Her gesture of thumping the floor of the cage with her front feet and emitting an explosive cough any time the male drew near her was an unmistakable warning, and to the male's credit it must be said that he was a gentleman and never forced matters. Over a period of weeks she got used to him and eventually the two animals became good friends, but they gave the zoo staff some anxious moments.

The cheetah or hunting leopard was once found in the plains areas over much of Africa and northward through Arabia and Persia to India, but it has now practically disappeared from the Asiatic parts of its range and is rapidly being reduced even in Africa. In many ways it is as un-catlike as a cat can be. Instead of the usual creep-and-pounce method of capturing prey, the cheetah is a runner, pursuing and overtaking its fleeing victims with sudden bursts of speed. While it can maintain great speed for only short distances, the swiftest antelope is usually pulled down within a few hundred yards. The cheetah is reputed to be the swiftest land mammal and has been reported as capable of speeds up to 70 miles an hour. Three checks with a stopwatch on a cheetah pursuing a mechanical hare on a dog racetrack, however, gave only 44 miles an hour, but this is still faster than similarly accurate records for any other mammal, including the racehorse and the greyhound. The fact that the cheetah's claws cannot be fully retracted, as in other cats, gives it the advantage of traction.

The combination of superlative speed, predatory habits, and curiously amenable nature makes the cheetah particularly adaptable for training for sport, and since early times it has been used by man for controlled hunting. This sport was highly developed in India on much the same lines as falconry,

*The cheetah, or hunting leopard of Africa and Asia,
has been called "the mildest cat" and is peculiarly
adaptable to training for hunting antelopes.
It is reputed to be the swiftest land mammal.*

and it still continues, although the number of cheetahs has so declined that suitable animals have to be imported from Africa. Just as wild-caught birds are preferred to hand-reared ones by the falconer, so is the cheetah with natural hunting experience esteemed by the trainer. A hand-reared hawk can be given partial liberty for "hacking" and then be returned to its block, but this is a practice hardly suitable for a cheetah. Once trained, the hooded animal is taken to the field on bullock cart or jeep and unhooded and released when game is sighted. The blackbuck, of course, is the favorite object of pursuit in India.

Despite the long captivity history of the cheetah, it appears that up to 1956 no instance of breeding in captivity had been recorded. In that year a litter of three young was born in the Philadelphia Zoological Garden. In anticipation of the birth the male had been removed from the cage and a covered nest box provided, but the female refused to enter the box and gave birth on the bare floor of the cage. She killed one cub almost immediately, so the other two were removed and bottle-fed, but they died three days later. The following year the female produced two more cubs and cared for them well for two weeks. Then, excited by noisy school children, she began carrying them about. They were taken away from her and rearing continued by hand, but they died of panleuco-penia at the age of three months.

Subsequently, two cheetah cubs were born in the Tierpark Krefeld in Germany and were successfully reared by hand, apparently the first successful captivity record. Two cubs were also born in the Oklahoma City Zoo but did not survive. Why the cheetah, "the mildest cat," is so reluctant to breed in captivity is still a mystery.

THE SEA LIONS

The oval seal lion pool that so frankly punctuates the long grassy sweep of Baird Court in the New York Zoological Park was built and stocked with California sea lions early in 1905. Almost immediately a pup was born to one of the inmates, and quite as promptly we lost it. One day it was seen struggling in the deep water and it drowned before we could rescue it. As later experience showed, this disaster was most unusual, as no sea lion here has since allowed her offspring to drown. The incident, however, has colored our thinking and our practice ever since. Zoo people have long memories when the welfare of their animals is concerned.

*Sea lion pups are unable to swim when they are very
young and are carefully guarded by their mothers.*

Actually, another birth did not occur among our sea lions
until almost exactly thirty years after the 1905 tragedy. Ex-
pecting the birth, we drained the pool except for a shallow
pond at one end, and here the mother and baby were confined. Within a month the pup had learned to swim and, with
its mother, was given the full liberty of the pool.

The first of a new series of births took place in the mid-
1940's when a female pup was born, this time in the rock
shelter along the north side of the pool. With 1905 in mind,
we placed low barricades at the entrances, to prevent the
youngster from falling into the pool. But after two days the
mother took the baby by the nape and swam with it to one of
the two concrete islands in the 88 by 44–foot pool, and tossed
it out of the water onto the concrete. Homelife on the island
then became routine, enlivened by the pup's frequent tum-

bles into the water. The resounding splashes drew the instant response of the mother and, rather surprisingly, of the father as well. Often the male was the first to arrive, but he never took what seemed to us the obviously required action of plucking the baby out of the deep water. His anxiety was always of short duration, but the mother never failed to arrive in a flash and to toss the youngster to safety. The pup, known as Benny, made no progress in learning to swim until she was about six weeks old. Her first experiences were gained by flopping up and down a series of stone steps that entered the water from the island. It took her two weeks to become proficient, so that she was eight weeks old before she could qualify as expert.

Benny endeared herself to staff and visitors alike by her playfulness, but it must be admitted that she was backward in some respects. Not only was she slow in learning to swim, but she was 17 months old before she was weaned, a period that certainly exceeds the normal expectation. We had high hopes for her as a lively member of the sea lion colony but when she was just under three years old she swallowed a small glass bottle that somehow was broken after ingestion, to her detriment.

Following the birth of Benny, her mother, Wendy, produced an almost annual succession of pups for several years, all but one of which, prematurely born, were safely reared with much less anxiety, it must be said, on our part. Most of them were much more precocious than Benny had been, and one was seen swimming and "porpoising" freely when only 13 days old. Under our pool conditions it was almost impossible for young sea lions to learn to eat, for at feeding time, when their keeper stands at the side of the pool and tosses fish rapidly one after another, they could only flee in the opposite direction to avoid being crushed in the mad rush of the adult animals toward the spots where the fish were landing. Therefore, when they reached an age between 10 and 11 months, we removed them to a smaller pool where, in solitude, they were taught to feed. If another pup was expected, they were not returned until after its birth. On being returned to the pool the older pup would usually attempt to nurse but was invariably rebuffed by its mother.

The California sea lion, ranging along the Pacific coast of North America from British Columbia to Mexico, is the species most commonly seen in zoological gardens. Noisy, active and graceful, it not only ranks among the most attractive of zoo exhibits but is capable of being trained for the performance of tricks involving balancing or leaping skills. Adult males may reach a weight of 600 pounds or even more —one in the New York Aquarium reached 620 pounds. Females are considerably smaller; our largest female was only 190 pounds, although one weighing 200 pounds is on record.

Marine animals though they are, salt water does not seem to be necessary to them. We keep them in fresh water only. Nor is temperature critical for these animals that are generally supplied to us from the warm waters of the southern part of their range. It is true that they generally stay in their unheated rock shelter in periods of severe cold, particularly if heavy ice forms on the pool, but on many a winter day visitors hasten from one heated building to another past the sea lion pool where the animals are diving and porpoising among floating cakes of ice.

THE WALRUSES

Not surprisingly, walruses are prize exhibits in any zoological park fortunate enough to obtain them—and to keep them. The males grow to enormous size; apparently between two and three thousand pounds. Practically confined within the Arctic Circle, they are found around the world in two forms, the Atlantic walrus and the Pacific walrus, although externally there is not much to choose between them. Both sexes have tusks, although the males' are the larger, and both young and adults have long, stiff bristles on the upper lip.

Our first walrus, which was also the first specimen to be received by an American zoological garden, came in 1902 as a gift of the Peary Arctic Club, but it lived for less than a month, the cause of death being reported as enteritis. In those days the complexities of walrus feeding were not understood and, indeed, there is still much to learn.

Our most recent experience with the walrus was certainly encouraging, if not entirely successful. This involved a young male named Herbert. On arrival by air, Herbert was in excellent health and condition, except for the usual skin lesions, and weighed 240 pounds. We were unable to learn the exact feeding schedule to which Herbert had been accustomed in Copenhagen, but on the basis of successful diets in the Copenhagen Zoo we improvised a formula consisting of undiluted evaporated milk, Mazola oil, and cod-liver oil.

At first Herbert refused steadfastly to take this thick, viscous mixture, his first meal here consisting of the flesh of mackerel, sucked cleanly from the bones, which were left intact. The lip bristles surely have a function here, for without their help a walrus could hardly remove the flesh so neatly. Anxious to get him onto our mixture, we added finely ground mackerel to the formula and presented the result to him in a large, deep pan, in which a few pieces of whole mackerel were floating. Herbert promptly reduced these bits to skeletons and having tasted the liquid inadvertently, proceeded to suck the pan clean. Twice daily thereafter, he consumed a quart of the mixture and, at first, 5 pounds of mackerel. As his size increased, the quantity of mixture remained the same but gradual increases in the fish were made, so that when he reached a final weight of 958 pounds he was getting 40 pounds of mackerel, which he still drew from the bones, and the usual 2 quarts of mixture daily in two meals.

Herbert spent his entire time with us in fresh water, and the skin lesions that were so apparent when he came to us soon healed without special treatment. During cold weather he slept in the water, usually with his head resting above it on the rock ledge. We thought that he might suffer from the direct rays of the sun in summer and stretched an awning across one end of his inclosure, but he never used it. When he had sunned himself sufficiently on the bank, he simply slid into the water to cool off.

Weighing the young Atlantic walrus "Herbert."

Two female harbor seals shared his pool and while he maintained friendly relations, there was never any great intimacy between the animals, for the wary seals skillfully avoided Herbert's constant efforts to clasp and hold them playfully in his front flippers. Herbert died less than two years after he came to us, as the result of an impaction caused by a rubber ball he had swallowed. Like most young walruses, he was engagingly friendly and his death was widely publicized and deplored, even in editorials in the New York newspapers.

Some interesting notes on the water preferences of captive walruses developed from observations made on the behavior

*"Ookie," a young Pacific walrus,
could easily scale this fence.*

of Olaf and Karen, two young Atlantic walruses received at the New York Aquarium. Kept at first in a small outdoor freshwater pool, the young animals seemed quite content. When sea water was substituted, however, the walruses managed to leave their inclosure and find their way to another basin filled with fresh water. A short time later, sea water was again run in, and again the walruses made nuisances of themselves by wandering. By this time the large tank that was to be their permanent home had been completed and was filled with sea water, and the animals were transferred to it. But they showed signs of distress by refusing food, and it was not until they were seen trying to catch raindrops in their open mouths that the cause of the trouble was suspected. When a hose was hastily connected to a fresh-water inlet and run into the pool, both young walruses drank from it heavily and almost immediately resumed feeding.

THE LEOPARD SEAL
AND THE ELEPHANT SEAL

Some years ago a British zoologist brought back from Antarctica a motion picture of the life of the Adelie penguin. Its briefest scene, yet one of the most dramatic, showed the dark, bobbing forms of penguins among floating ice in a rough sea. Suddenly a dark head rises from the water, a penguin is lifted and brandished a few times, the head sinks and the stripped skin of the bird floats away. A leopard seal had snatched a hasty meal, literally shucking the carcass out of its skin.

The leopard seal is found here and there in the seas about Antarctica, and perhaps because of the remoteness of its habitat, it has seldom been seen in captivity. A long, slender-bodied animal, the adult males reach a length of 12 feet. In addition to the usual seal diet of fish, squids, and crustaceans, it also takes warm-blooded prey, including young seals of other species, penguins, other sea birds, and even carrion. A stranded male taken in 1870 near Sydney, Australia, is reported to have swallowed a full-grown platypus!

My own experience with this species is limited to the viewing of a living specimen at Hagenbeck's Tierpark at Stellingen in Germany in 1912. This animal was appropriately shown at the front of a series of moated inclosures, with king penguins above and directly behind. As the leopard seal floated quietly in the water, it looked straight at me, giving a definite impression of sinister invitation, enhanced by the immense extent of its jaws. It was a chilling, unforgettable experience that I was never able to analyze fully until I read the lively description given by the scientific director of the Zoological Society of London in which he ascribes to the leopard seal a reptilian appearance. That is exactly the point: no seal should suggest a crocodile!

Apparently the leopard seal's wild habits carry over into captivity, for when a specimen was kept in the Taronga Zoo in Sydney it decimated a flock of ducks, to the mystification of the attendants. A leopard seal lived at Stellingen a little over 2 years and at death was 10½ feet long and weighed 957 pounds.

Largest of all the seals is the elephant seal, the southern species of which is found at scattered points in the Subantarctic area and the northern off the coasts of California and Baja California, with the largest concentration at Guadalupe Island. Fortunately its ferocity is not in proportion to its size, for males of the northern species grow to a length of 22 feet and a weight of 8,000 pounds, and the southern species only slightly smaller. The name comes from the enormous, distensible, pendulous trunk of the adult males.

Any zoo would, of course, be overjoyed to have a huge male elephant seal, but it must be admitted that it would not be an esthetic sight. The thin, coarse hair is shed annually, the epidermis stripping off in patches. This results in a most unsightly appearance in captive animals, since the bits of hair-covered skin are inclined to dry and curl at the edges, even though the animals have free access to water. Molting elephant seals on South Georgia Island indulge in mud baths, a custom that certainly should be of great aid in the sloughing off of skin but unfortunately not easily adaptable for use in a zoological garden.

In Europe, the Hagenbecks, the German animal dealers, pioneered in the importation of elephant seals and between the two world wars brought about 24 specimens to Stellingen. The first adult male to arrive was the famous Goliath, who came in 1926. He was then 16½ feet long and weighed 4,821 pounds. The maximum quantity of fish Goliath ate in one day

Even young elephant seals are impressively large. When full-grown, the northern elephant seal may be twenty-two feet long and weigh eight thousand pounds.

was 385 pounds, given at intervals during "performances," but when he was in top condition, 80–100 pounds a day were enough to keep him going. Goliath was sold to Ringling Brothers around 1928 and spent several years traveling with the circus, where his spectacular appearances are unlikely to be forgotten by anyone fortunate enough to have seen them.

Most of the larger zoological gardens have had elephant seals at one time or another—the gigantic Roland in the Berlin Zoological Gardens, Jonah and Freya in the St. Louis Zoological Park, Nixe at Stellingen, and many others. Nixe, incidentally, lived for about 15 years, paralleling the longevity of another specimen in a European circus.

Most of the experience in this country has been with the more readily obtainable northern elephant seal, and the San Diego Zoological Garden especially has exhibited them frequently, since they may be captured nearby. Some of the males in particular caused the feeding difficulties common with large specimens when newly captured and were forceably fed after being wrapped in canvas sheets. This continued for as long as six months, in one instance, before the animal began to accept fish voluntarily. One of San Diego's males was very large, measuring 16½ feet and weighing nearly 5,000 pounds on arrival.

It is sad, and a little ridiculous, to relate that a 3,800-pound male in the Chicago Zoological Park died of an intestinal impaction caused by swallowing peanuts thrown to him by visitors! As a matter of fact, our own first elephant seals came to us as a result of a rather discreditable circumstance. Acquired for the New York Aquarium's collection, they so persistently annoyed a young walrus that they were transferred to the zoological park. Both were young, the male weighing only 320 pounds and the female 625 pounds. We had no suitable place for them except the pool already occupied by an established group of sea lions, and these vigorous and active animals so persecuted the male that he refused food and had to be removed to a smaller unit, where he quickly settled down in the company of an inoffensive gray seal. The larger female elephant seal was better equipped to hold her own and was able to establish herself among the sea lions with no further trouble. During their first winter temperatures went down to 5° above zero, but both elephant seals, which had come to us from comparatively mild California waters, remained in the water, usually below the surface, coming up to breathe at intervals through breaks in the ice made by themselves or other occupants of the pools. They did not live to attain the immense size of which the species is capable, and a really gigantic elephant seal is something we still hope for.

THE ELEPHANTS

No matter how familiar one may become with the over-powering bulk of elephants, the marvel never ends. I once heard an old, very experienced keeper, working with one of his charges, say to it, "Move over, you wonderful, blasted lummox!" And one of the many wonderful things about these great lummoxes is that, ordered by a puny keeper to move over, they ponderously and obligingly do so.

The elephants are the largest living land mammals of the world; that much is incontrovertible. It is only when one comes to examine maximum height measurements and weights that the ground is not always firm—and for obvious reasons. Measurements of animals killed in the field are useful for comparison but certainly do not represent true standing heights. Weights of such specimens cannot usually be obtained with any degree of accuracy, and even where captive elephants are concerned, walk-on scales are not always available, and when they are, large animals usually cannot safely be led to them. For that matter, neither can small elephants—at least, if they have made up their minds that wobbly platform scales are unsafe. I well remember how it took the combined efforts of all the keeper, clerical, and curatorial help that could be mustered to push a very young, but very timorous, elephant onto the truck-weighing scales at the zoological park on one occasion.

Measurements of height taken from ground to ground across the shoulders do not give true heights when divided by 2, and computations of twice the circumference of a forefoot are equally inaccurate. An adjustable arm, attached at right angles to an upright, applied across the shoulders of a standing elephant, does give the height accurately—that is, if the elephant will stand quietly while such a strange contraption is placed alongside it. Not all will! We have sometimes been successful with this method, and at other times have had recourse to an engineer with surveying instruments.

Regarding the Indian elephant, authorities say that most males do not exceed 9 feet, females 8 feet, but considerably greater heights are possible; one male is said to have stood 10 feet, 7½ inches and there are estimates of others up to nearly 12 feet high. There can be no doubt, however, about the authenticity of statistics concerning Bolivar, the large male Indian elephant kept in the Philadelphia Zoological Garden from 1888, when he was thought to be 27 years old, until his death in 1908. Bolivar's shoulder height was 10 feet and his weight approximately 12,000 pounds.

Accredited data concerning African elephants are even harder to come by. It is unfortunate that the famous Jumbo, undoubtedly a very large animal, did not leave a really definite height record. Jumbo, while still quite young, was received in exchange from the Jardin des Plantes in Paris, by the Zoological Gardens of London, on June 26, 1865, and a few months later was found to stand 5 feet, 6 inches. Jumbo proved to be extremely tractable and was used for some years as a riding animal. He gradually became less reliable as he neared maturity and by 1881 could be approached only by his keeper, so he was sold to P. T. Barnum for the then great sum of £2,000. Jumbo was crated and shipped to America, arriving in March, 1882. His age at that time was said to be about 21 years and his height as nearly 11 feet. Cleverly publicized, Jumbo achieved great fame with the Barnum and Bailey Circus until he was struck and killed by a railway engine at St. Thomas, in Canada, on September 15, 1885. His weight at death is usually given as 6½ tons or 13,000 pounds. His height, however, will never be accurately known, for his owner is said not to have allowed it to be measured. Dr. William T. Hornaday, later to become the first director of the New York Zoological Park, made futile efforts to obtain permission to make the measurement, invariably getting what today would be called a run-around. Nevertheless, he later received a report of a circus pole-jumper who had, as if casually, stood his pole alongside Jumbo, while marking the shoulder height with his eye. This "measurement" showed Jumbo to be 10 feet, 9 inches at the shoulder, and was probably the most nearly accurate figure ever made known.

Jumbo's closest rival among captive bush elephants was Khartoum, who lived in the New York Zoological Park from 1907 to 1931. His weight at death was 10,390 pounds and his height 10 feet, 10 inches. This measurement was made after Khartoum's death and while he was in a prone position, so it somewhat exceeds his greatest measured standing height, which was 10 feet, 8½ inches.

*Friendly play between a female African bush elephant
on the left and a male African forest
elephant on the right.*

Probably the greatest true weight ever recorded for an elephant is the 14,641 pounds for a huge bull dismembered and
weighed in parts by Dr. George Crile, the endocrinologist,
while on an expedition in Africa.

The living elephants are, in reality, a relict group, the present survivors of a great assemblage of related but now extinct
forms once abundant throughout the Northern Hemisphere
and even the polar region. As we know them today, the elephants exist only in southern Asia and in Africa south of the
Sahara.

We keep African and Asiatic elephants in adjacent yards
and the differences between them are readily distinguishable
when they are seen side by side. The ears are their most obvious characteristic. Comparatively small and roughly triangular in shape in the Asiatic, they are huge in the African bush
elephant. "Africa—a big continent. African elephant—big
ears" is the way one visitor told me he distinguishes them.

Oddly enough, the Asiatic elephant fairly frequently shows
depigmentation in the form of whitish or pinkish spots, a
color deviation that is rare or almost absent in African elephants. Albino elephants sometimes occur in Asia and are

regarded with reverence or even held sacred. No work is required of such animals, a situation that seems to have given rise to the opprobrious term "white elephant."

Beyond their huge size and immense strength, the elephants' most remarkable character is the trunk. This unique organ, actually a prolongation of the nose and upper lip, performs a variety of essential functions. It carries the nostrils at its tip, so it serves basically in breathing and smelling. It also acts as a prehensile hand in securing food and transferring it to the mouth; water is drawn part way up its length and then discharged into the throat. Capable of feats requiring both power and mobility, the trunk is extremely delicate and sensitive; I once saw a great African bull place the tip of his trunk over a nest of young robins on the fence of his inclosure, test it cautiously, and gently withdraw, leaving the fledglings to be safely reared.

The Asiatic elephant is able to make a hollow, resonant sound by tapping "back-handed" on a hard surface with its trunk, the tip of which has been turned upward. This appears to be a warning or alarm signal, heeded at once by others of the species within hearing. I have never seen this device used by African elephants.

Everything about the elephant seems outsize and of special interest. This is certainly true of the teeth. Most prominent, of course, are the tusks, which are in reality a single pair of upper incisors, typically greatly elongated in males alone of the Asiatic elephant and in both sexes of the African forms. Actually, the female Asiatic does possess tusks but these are usually not sufficiently developed to protrude beyond the lips.

Elephant tusks, as the chief source of ivory, have been so sought after throughout the centuries that the elephant population has been reduced in all parts of its range and in some areas has been virtually exterminated. While the tusks of Asiatic elephants do not equal the dimensions of those of African origin, they nevertheless may be very large. There is a record of a pair that measured 8 feet, 9 inches, and 8 feet, 6 inches and weighed 162 and 160 pounds, respectively. The longest African tusks seem to be a pair in the National Collection of Heads and Horns in the New York Zoological Park. These measure 11 feet, 5½ inches and exactly 11 feet, respectively, and together weigh 293 pounds. A single tusk only 10 feet, 2½ inches long but weighing 226½ pounds is owned by the British Museum (Natural History).

Besides the tusks, the elephant's only teeth are the molars, a total of 24, 6 in each jaw, appearing during an elephant's lifetime. Only one in each series, perhaps with portions of its predecessor and successor, is functional at the same time, and as it wears down it is pushed out at the front by another moving forward from the rear. The last tooth to come into use is usually the largest and may be nearly 12 inches long. These immensely heavy teeth, strongly cross-ridged, are a grinding apparatus of great crushing power so that an elephant can grind up tree branches of some thickness with ease.

General opinion today is that there are 4 races of elephant in Africa—3 of the so-called bush type, and the smaller forest elephant. At one time there was supposed to be a pygmy elephant in Africa—in fact, a male named Congo that came to the New York Zoological Park in 1905 was scientifically described as the type of this supposed subspecies. The designation no longer holds true, and while the possible occurrence of pygmy elephants is still in dispute, there seems no doubt that if chance dwarf individuals or groups do exist, they are still just forest elephants.

I have good reason to remember Congo, our first forest elephant. He was an ill-natured little creature, fully controllable only by Alice, a steady Indian cow elephant who adopted him. He once nearly succeeded in impaling me on his slender tusks, between which, at the last moment, I managed to slip. He always seemed to us to be afflicted with arthritis and his keepers in the early days of the century went to the length of fitting him with tightly laced leather ankle supports. Nothing did any good, however, and eventually it was decided that the most humane thing was to put him to sleep. Carl Akeley, the African explorer and hunter, was called on to end Congo's life with a rifle bullet. Congo's remains now repose in the American Museum of Natural History in New York.

The elephants' gait is what is known as the amble or pace. In this action the feet on one side move forward together, alternating with those of the other, so that the trot or the gallop is impossible. This does not mean, of course, that the elephant is not able to move with speed, at least for short distances. One frightened elephant, timed with a stopwatch, covered 120 yards at the rate of 24 miles an hour.

The matter of sleep in elephants has long been in dispute. A number of investigators have shown that while elephants of all ages may sleep for short periods while standing, all but aged or infirm animals do lie down, usually flat on one side and often with a pile of hay under head or body. This usually occurs after midnight, and sleep may continue for two or three hours in adult elephants, but longer in immatures. It is difficult to observe elephants asleep in the zoo, since they are not accustomed to noise at night, but their sleep has been readily studied and even photographed in circus and working animals. Elderly elephants appear to sleep only while standing, perhaps because of the difficulty of rising from a prone position. Even though elephants sleep so lightly that they may not easily be approached while lying down, flanks stained by dung betray them in the morning.

Mature male Asiatic elephants in captivity are subject to periods of restless irritability, when they are likely to damage their surroundings and become dangerous even to keepers with whom they have long been on good terms. The usual sign of this condition is enlargement of the temporal glands, which lie between the ear and the eye, followed by discharge of a black, oily fluid. This state is known as "must" or "musth" and is said to occur, but rarely, in females also. Close observation over many centuries has made "must" well known wherever the Asiatic elephant is kept as a work animal. Since, of course, the daily work routine of males in "must" is disrupted, many remedies for shortening its course have been devised. The most effective of these seems to be the limiting of food to reduced amounts of greens only and the usual precautions of restraint. Whether or not "must" has a sexual significance seems not to have been definitely determined.

The male African elephant in captivity is likely to become obstreperous, even dangerous, when adult, but it does not seem to undergo periods of "must" like those of the Indian elephant. Khartoum, the African bush elephant mentioned above, was high spirited and destructive but never showed animosity toward his keepers. On the other hand, little Zangelima, a forest elephant, became so increasingly antagonistic and dangerous that it finally became necessary to destroy him. Neither of these animals showed evidence of a periodic disturbance that could properly be described as "must."

As might be expected, housekeeping arrangements for elephants in a zoological garden are not simple; docile as the animals usually are, the off-chance that on some occasion they may not be docile has to be considered, and there is always the problem of their enormous strength.

Steel bars can perhaps best be described as the traditional way of restraining elephants and that feature was built into the New York Zoological Park's elephant house when it was built in 1908. Fronts of the four compartments for elephants are 2-inch round steel bars, 10 feet high, set in concrete at the base on 20-inch centers and joined at the top by two heavy steel cross-members. Keepers can enter the compartments for cleaning only by slipping between the bars.

The alternative to bars is the indoor moat, used by several zoos in this country and abroad. An elephant will not ordinarily attempt to cross a moat 6 feet wide and 5 feet deep, and the minimum safety span between the animal and the public is 10 feet—which of course means that the moat has to be both wide and deep and a guardrail set well back in the public space. Such a moat must have its sides perpendicular or sloped at a sharp angle, so that, if an elephant gets into it, either by accident or design, it is by no means easy to get it out—and the animal may well be injured in the process. Obviously, what was needed was some way to keep the elephant from getting into the moat, and at one time it was the practice to set rows of iron spikes across the rim of the moat. Frequent injuries to elephants' feet, however, to say nothing of the implications obvious to the public, have brought this method into disrepute. Consequently, some of the newest buildings for elephants, rhinoceroses, and hippopotamuses, while employing moats to restrain the latter two, still use bars for elephants.

Out-of-doors elephant inclosures present the same conflict of bars or fences versus moats, but the moats have the better of it in modern installations.

We gave up iron-barred yards many years ago, but before we did so we had several manifestations of the strength of an elephant. The yard allotted to Khartoum, the big African bull, was fenced with heavy steel beams securely, or so we thought, welded together. Yet Khartoum, in merely an exploratory mood, managed to so twist and batter this fence that it had to be replaced with steel rails double the original strength—studded, it must be admitted, with stout, blunt spikes. Only then were Khartoum's destructive efforts ended.

These unsightly barriers were finally replaced with a stone wall. The central portion of the big outside yard was left at a level slightly higher than the public viewing area, but the yard slopes down to the inner base of the stone wall at a well-graduated angle. The result is a half-moat, or ha-ha, that has the effect of increasing the safety margin between the elephant and the public, because when it extends its trunk in begging, a prerogative of all zoo elephants, the animal is either standing downhill facing forward, or sideways. Steeply as the ground breaks away before the wall, it by no means prevents the animal from going right up to the wall and it can walk back to level ground just as easily.

The wall of our yard for small elephants extends upward only 4½ feet from the bottom of the ha-ha. The first occupants of the yard were two female Asiatic elephants, estimated to be about 15 years old and standing approximately 7 feet, 10 inches at the shoulder. For a time all went well, but one morning I was informed by a keeper that one of the elephants had jumped over the wall and, while she had allowed herself to be led back, she was still greatly excited. Needless to say, I ran for the elephant house and arrived just in time to witness an extraordinary spectacle: an elephant jumping over a wall! When I reached the front of the inclosure, the animal, known as Cutie, was standing in the doorway of the inner stall, facing outward, with ears extended and trunk waving. Suddenly, with a loud trumpet, she charged directly at us at full speed. Without checking, she ran into the corner of the moat, turned sideways, and threw her right feet over the wall, so that her momentum, probably aided by a push against the slope with her left legs, carried her completely over. Once across, she stood trembling in the space between the outer face of the wall and the light guardrail, obviously not knowing what to do once the primary obstacle had been cleared. For the second time she walked quietly back to the gate leading to the inclosure, a keeper holding onto one of her ears. Cutie had no opportunity for further development of her project, for she and her companion, Dolly, were promptly transferred to the opposite inclosure, occupied by the steady old African cow, Sudana. Sudana lost no time, butting and trunk-slapping, in convincing the Asiatics that she was mistress.

The difference in temperaments among elephants was well illustrated by subsequent events. Dolly remained calm enough after the transfer but Cutie was still bent on exploration. In her new yard, the wall extended 6 feet, 6 inches upward from the moat bottom, and Cutie soon found that she could increase the reaching distance of her trunk—and thus her share of the food offered by the public—by putting her front feet on the wall. We were satisfied that she could not repeat her performance by jumping over such a high wall, but she was able to reach farther than we liked, so electrically charged cattle guard wires were installed near the top of the inner face of the wall. All the elephants of course investigated such a novelty and received mild shocks that sent them running off with loud shrieks and with much trunk-tapping on the part of the Asiatics. For two or three weeks they shunned the wall, but they gradually learned to avoid the wires, or even to ignore the shocks, and they eventually ripped the wires off the wall.

Put to a better purpose, Cutie's persistence would have been admirable. Exploration with her trunk informed her that the outer, or public, face of the stone wall rose only 2 feet above the grass plot between the wall and the guardrail. By getting her front feet over the wall and pressing them against the exposed outer surface, she was able to draw herself upward. When she had reached the point where only the toes of one hind foot remained on the ground and it was obvious she might very well succeed in getting over, we reluctantly resorted to a light cross-chain—right rear to left front—that ended the matter.

After a good many years of observation, it seems to me that there is a kind of symbiosis between elephants and the public —the one loves to beg, the other to give food. From time to time we have made more or less serious attempts to stop all feeding of the animals in the zoo by the public, but we have always had to make exceptions with the elephants. The elephants *will* beg, and the public *will* feed, if there is any way for reaching trunk and extended arms to make contact. On a busy day, when forty or fifty thousand visitors are in the Zoological Park, the amount of popcorn, food pellets, and other edible material offered and accepted must be large, but it seldom if ever affects the elephants' appetite for their more normal food.

Elephants in nature, as well as captive animals, consume great quantities of leafy and grassy foods where hay is not available, perhaps as much as 600–700 pounds daily. For zoo animals, fed largely on dried materials, the actual weight consumed is naturally much less. Sudana, when she had attained a standing height of 8 feet, 7 inches at the shoulder, required approximately 300 pounds of timothy hay daily, given at intervals. She set her own mealtimes, for she trumpeted loudly when the supply was low and "purred" softly when it was renewed. In addition to hay, Sudana received 16 quarts of mixed ground grains, containing minerals and salt, as well as four or five loaves of bread, three or four cabbages, and raw

Captive elephants are inveterate beggars. This is "Candy," a young Asiatic elephant.

white potatoes, apples, and carrots as treats. In summer, she ate green stalks of corn, grass, and leafy branches as they became available. All this bulk seemed to have no effect on her begging for popcorn.

Elephants consume water in great quantities, an animal of average size requiring somewhere between 35 and 50 gallons a day. We follow the usual custom of watering twice a day, morning and night, and never allow water to stand in the stalls, for then it serves only as a plaything, to be squirted on the animal's back—and perhaps, inadvertently, innocent bystanders.

Care of the skin of zoo elephants has always been a matter of some concern, since it may become dry and thickened, especially when the animals are confined indoors. At one time we followed the then usual practice of applying neat's-foot oil in liberal quantities, followed by a brisk scrubbing.

Today we rely entirely on the hose, each animal receiving a daily wash-down with warm water, accompanied by a thorough massage with a stiff broom. This treatment keeps the animal's skin in good condition and is apparently enjoyed, if the spectacle of a huge elephant squirming about on the floor of her stall, presenting different surfaces for the attention of the broom, is any indication.

Fairly extensive studies of the reproductive habits of Asiatic elephants have been made. Accumulated data indicate that males are usually sexually mature at 14–15 years of age, while females are usually 15–16 years old when their first calf is born. The gestation period—a matter about which there is widespread public curiosity, to judge by the telephone inquiries received by the zoo!—is usually 19–21 months. This may seem to be a considerable spread, but there are records of gestations as short as 17 months and as long as 24 months. Rather less is known about the reproduction of the African elephant, but such data as there are suggest that both sexes mature at 8–12 years, and that the gestation period is about the same as for the Asiatic elephant.

Births are usually of single young, although twins have been recorded. The newborn young are heavily coated with hair that gradually becomes reduced, although varying amounts persist through life. While the paired mammary glands are located well forward on the mother's chest, between the forelegs, the infant is nevertheless able to apply its mouth directly to the nipple, although it was once thought to use its trunk for the purpose. After several months the calf begins to eat green foliage and grass, although it probably continues nursing for about two years. Even after the youngster is entirely able to feed itself, it is still under parental supervision.

Births of elephants in zoological gardens are rare, although by no means unknown. The basic reason for the infrequency of zoo births is the difficulty of maintaining and controlling mature bulls under exhibition conditions without constant risk of injury of keepers or even the public. If quarters can be provided primarily for breeding purposes, with exhibition a secondary consideration, and the safety of keepers and public assured, there seem to be few reasons why elephants should not breed reasonably well in captivity. Two births of Asiatic elephants occurred in Europe in 1906 and in Buenos Aires in the same year. Since that time more than a dozen births have occurred in Europe.

The first elephant birth in this country appears to have been one that occurred on March 10, 1880, the mother being a cow named Hebe belonging to the Cooper and Bailey Circus, then wintering in Philadelphia. The calf is said to have reached the age of 25 years, when it was destroyed as a "killer." In subsequent years, five more live births are recorded among circus elephants, in addition to one still-born calf.

Births of elephants in zoological gardens in this country— at least of calves bred in the zoo—date only from 1962 when a male calf was born to Belle at the Washington Park Zoo in Portland, Oregon. The calf was reported to weigh 225 pounds at birth, and the gestation period was 20 months, 26 days. Within two years, a total of four calves were born to Indian elephant cows in the Portland Zoo, all sired by the bull Thonglaw. It is significant that all doors in the Portland elephant house are operated by hydraulic-electric control and that shifting of the animals is done with complete safety. Thonglaw is dangerous during his "must" periods but can always be brought back under control when this subsides.

As with most other phases of interest in elephants, the age to which they may live has long been subject to misunderstanding and exaggeration; there are many tales of elephants having lived well over 100 years. One Indian naturalist believes that elephants in the wild state may live as long as 120 years under good foraging conditions, though only to 70 years under less generous ones. There seems to be no practical way to determine this, however, and certainly authentic records of captive elephants, especially those kept in zoological gardens under presumably optimum conditions, indicate a maximum life span of far less than 100 years.

Our best record for an Asiatic elephant is that of Alice, received in 1908 and destroyed in 1943, after a span of 34 years, 11 months, 24 days in the New York Zoological Park. Alice is presumed to have been almost exactly 50 at death. For several of her later years she had been showing signs of senility and no longer lay down to sleep. Four times in her last 3 years she had gone down while out of doors and, being unable to rise, was set on her feet by a derrick. On the fifth occasion, her plight was deemed hopeless and she was quietly euthanized.

A search of all available authenticated records gives the palm for longevity to Jessie, an Asiatic elephant in the Taronga Zoological Park in Sydney. There is some doubt whether Jessie should be considered as 12 years old, or 20, on arrival at Taronga; in any event, she lived in that zoological garden for approximately 57 years, and whether she was 69 or 77 years old at her death, she was certainly the oldest elephant of which there is an acceptable record.

THE ZEBRAS

The admirable practice of exhibiting numerous animals of the same continent or geographic area together has led, quite naturally, to the creation of "African Plains" or "African Veldt" exhibits. Equally naturally, attempts are made to sprinkle the exhibits with large and striking animals, and among these few can equal the various species and races of the zebra. Unfortunately, the zebras are not always cooperative.

It is true that in the Detroit Zoological Park I have seen zebra mares living quietly in summer with elands, lechwe, and ostriches on an African Veldt, but the zebra stallion could not be so privileged and for much of the time was confined to winter quarters. In our own experience we have found zebra mares to live quietly with elands of both sexes, but on one occasion when circumstances compelled us to add aoudads to the group, the experiment assumed a different

The width and wide separation of the stripes on the thigh are characteristic of the mountain zebra of southern Africa.

aspect. The zebras hunted the small aoudads so persistently that only the superior agility of the latter saved them from destruction. After a very short trial we removed the zebras and after a cooling-off period tried returning them, one at a time, in the hope that at least one less aggressive individual might be found. But it was no use; even an elderly zebra, so afflicted with heaves that any movement seemed difficult, promptly proved herself more able than we had thought her.

At another time we tried a zebra stallion, aged barely 2 years, with a mixed group of antelopes and birds on our African Plains. For a time all went well, but eventually the stallion's interest became focused on a pair of wart hogs that he soon confined to a burrow they had dug in a bank above a pool. When the zebra finally crushed through the roof of the burrow, the experiment came to an end.

Once abundant on the plains and in lightly wooded areas of Africa, east and south of the Sahara and the great forests of the west, from Ethiopia and Angola to the Cape, it is a regrettable fact that zebras are rapidly losing ground. Some forms have disappeared entirely within recent times, while others are perilously close to extinction. Zoologists generally agree that the living zebras fall into three species: Grevy's of southern Ethiopia and northern Kenya; the Burchell group that includes three races and is found in eastern, central, and western Africa, as far north as Angola; and the mountain zebras, from the south. There was once a peculiar zebra known as the quagga, which was striped in brown and white on the head, neck, and forepart of the body, with the remainder of the upper parts solid brown and the belly, legs, and tail white, but it is now totally extinct. The last individual died in a European zoo in 1872, 1875, or 1883—accounts vary —and in the wild they had apparently been exterminated long before even the earliest of these dates.

Many tropical or subtropical animals are surprisingly resistent to cold, but in the latitude of New York the zebras cause borderline problems. We maintain a temperature of about 60° F. in our zebra house during the winter months, but turn the animals out into their runs daily, unless the weather is too severe or temperatures are below the freezing point—zebras are not clever on icy footing.

The public's urge to feed the animals in the zoo extends, of course, to the zebras, for they are so reminiscent of gentle horses in costume-party disguise. Because of this belief that no creature as beautiful as a zebra could be ready to bite and even maim the hand that offers it a tidbit, we cover the 7-foot, iron-bar fence of the zebra inclosures with 2-inch mesh. It is only too well known, to zoo people, at least, that zebra stallions and even the occasional mare, of any species, may be savagely aggressive and dangerous. In this category I recall a Grevy and a Grant (one of the Burchell group) that would attack man or beast at any opportunity.

The 7-foot height of all the fences in our zebra yards has so far proved adequate, as zebras are not famed as high jumpers.

There is a record, however, of one male zebra, startled by a falling branch, that cleared a fence 4 feet, 10 inches high from a standing start.

Zebras breed well in captivity, and there are also many records of crosses between zebras and horses and zebras and asses. They are long-lived in general; the best record for a zebra is that of a Chapman's mare in the Zoological Gardens of Basel—28 years, 1 month, 24 days.

THE TAPIRS

As far as I am able to determine, no mountain, or woolly, tapir, had ever been exhibited alive until a combination of circumstances produced one for us in late 1950. Indeed, for a long time the very existence of the animal was doubted—so doubted, in fact, that when, in 1929, the late Dr. William T. Hornaday wrote an article in the New York Zoological Society's magazine, "Tapirs, So Far As Known," he made no mention of the mountain, Andean, woolly, hairy, or Roulin's tapir—all names which at one time or another have been applied to the animal.

As late as 1950, while we had none of Dr. Hornaday's doubts about the actual existence of the animal, we were not entirely prepared to believe that we were going to have the good fortune actually to exhibit one. Too many times we had been offered rarities that turned out to be something entirely different.

As it happened, the mountain tapir was one of the animals on our "most wanted" list when we sent the animal collectors, Charles and Emy Cordier, to Ecuador in 1949–50. Charles faithfully searched and even advertised in the Quito newspapers, and just as he was on the point of returning to New York with an otherwise fabulous collection of mammals and birds, he heard of a tapir living as a village pet in a tiny hamlet some 60 miles east of Quito, at an elevation of about 6,000 feet. It was too late for Charles to travel so far and when he attempted to work through an intermediary, the price of the animal became so astronomical that he abruptly dropped the negotiations.

Among the friends who went to the flying field to see him off when he returned to New York was a Swedish explorer resident in Quito and as a more-or-less afterthought Charles told him about the tapir he had been offered—which he was convinced was one of the rare mountain species. Not long afterward, his friend went to the village and by direct negotiation got the price down to a reasonable level. He thereupon asked us by letter if we wanted to buy the animal.

Its name was Panchita, it seemed, and it was a village pet, wandering where it would, and petted by everybody. Still a baby, Panchita was wearing the striped and spotted coat of tapir infancy.

The letter communicating this information gave assurance that this was indeed a mountain tapir, but at that time so little was known about the animal and descriptions of its external characteristics were so confusing that we felt a mistake might have been made in all innocence.

While we hesitated, another letter arrived from Quito, this time inclosing three tiny photographs, each one scarcely larger than a postage stamp. They resolved our indecision; Panchita might not be a mountain tapir, but she was certainly unlike any other tapir we knew about. We ordered her by cable, and Panchita was soon on her way by air.

If there had been any doubts left, they ended when we opened the crate in the elephant house. Her body was clothed with dense, matted hair, blackish-brown and actually kinky (we remembered the alternative common name of woolly tapir), and there was a broad white band nearly an inch wide entirely around both lips, as well as a half-inch ring of bare whitish skin above the toes all around. There were several other unmistakable diagnostic characters, but they were not needed; we were seeing a mountain tapir alive for the first time.

The mountain tapir is certainly the rarest but by no means the most spectacular of the tapirs; that distinction belongs to

Keeper J. Coder feeds a baby Baird's tapir.

the Malay or saddle-back tapir that ranges from Sumatra
northward through the Malay Peninsula to the borders of
Burma and Thailand. Like other tapirs, it is blackish or blackish-brown in general, but has a blanket of white or grayish-
white extending over the back from shoulders to hips, rather
as if it had fallen over backwards into a tub of whitewash.

All the other tapirs of the world are from Central and South
America—Baird's tapir the largest, the Brazilian tapir next,
and the mountain tapir the smallest. The Malay tapir perhaps
slightly exceeds Baird's in size; we have had Malays that

weighed as much as 690 pounds, and the San Diego Zoo weighed a female at 750 pounds.

Wherever they are found, tapirs frequent heavy jungle, often swampy or close to streams or lakes, either in tropical lowlands or high in mountainous regions. They swim well and take freely to water when pressed by enemies. Their natural food is low-growing forest vegetation and various fallen fruits. Night is their usual time for roaming in search of food and both in the wild and in captivity they pass much of the day in sleep, although they quickly become alert, even in bright sunshine, if they are disturbed.

Most references to them in the wild describe them as harmless, defenseless, and slow-moving, but our experience is that an aroused tapir can be a serious antagonist, able to use its teeth effectively and capable of surprising speed and agility. Hand-reared specimens usually remain tame and gentle and are often kept as pets around native villages—as our young Panchita was—but they are still subject to occasional moods best described as "tantrums," when they are anything but tame and gentle. I recall one male Brazilian tapir that was much given to tantrums. His bathing tank in the elephant house was fronted by ½-inch plate glass for better visibility, and in one of his frenzies he broke the glass, as well as a replacement sheet we hurriedly set in place. It was not until we substituted shock-resistant laminated Herculite glass that his charges were foiled.

THE RHINOCEROSES

Since the rhinoceroses are rivaled only by the hippopotamus as second to the elephants among the greatest of the living land mammals of the world, it might be expected that their exhibition would present many difficult problems. Actually, if normal safety devices are installed and precautions taken, the problems are solved readily enough.

Sometimes a surprisingly simple device is effective. Accom-

The African black rhinoceros, an object of great curiosity to Asiatic elephants in an adjoining yard in the New York Zoological Park, is a dark brownish-gray rather than black.

modations for rhinoceroses in the New York Zoological Park are, traditionally, in the elephant house, where winter heat is available. There are only two stalls, each 24 feet square and each fronted with 2½-inch steel bars on 20-inch centers. These bare stalls, just as laid out in the original construction, were far from convenient and certainly lacked provision for safety, since no area was provided where a rhinoceros could be shifted for safety when the keeper had to enter the stall for cleaning. As a first step toward improving this condition, a concrete wall only 38 inches high was built from front to back across the compartment occupied by a particularly obstreperous black rhinoceros bull. Spaces 5 feet wide were left at front and back, so that the low wall only partly divided the stall and the animal could circulate freely. A heavy chain was so arranged that it could be drawn across the forward gap, hanging loosely about 12 inches above the floor. Such a barrier would not seem to be impassable to a big and determined animal—but it was. For 10 winters the rhinoceros struggled daily with this apparently slight obstruction and never once succeeded in crossing it, so that servicing could be carried out on one side after the other. Nowadays hydraulically operated doors under remote control, and a former storeroom doing duty as a shifting cage, insure complete safety for the keeper. Nevertheless, he worked safely for all those years with only the chain between himself and the animal. He has, however, been heard to admit that he feels a *little* easier when the rhinoceros is shut up in the shifting cage behind a steel door!

Rhinoceroses today are relics of a once numerous and widely distributed group, and of the five living species, three are native to Asia and two to Africa. It is their great misfortune that all possess horns, for there is a persistent belief in Oriental countries that ground-up rhinoceros horn is an aphrodisiac, and as a consequence, rhinoceroses of any and all species have been implacably hunted, several forms almost to extinction.

In all three species of rhinoceros found in Asia, the thick skin is arranged in folds with thinner and more pliable areas lying between them, giving an armor-plated effect—indeed, medieval artists depicted the animal with bolt heads protruding from its "armor." This warty skin does depend on its thickness for its defensive value but it is far from being bulletproof, as was once believed.

The plates of the skin reach their greatest development in the Indian rhinoceros. Both sexes carry a single horn, and the record specimen is 24 inches long, a respectable length but by no means the greatest. With a shoulder height of 5 feet to 5 feet, 9 inches, and a weight frequently quoted as about 4,000 pounds, the Indian is, of course, a very large and powerful animal—but again, not the largest. The white, or square-lipped, rhinoceros of Africa has that distinction.

Asia has two other rhinoceroses, the Javan or lesser one-horned, now almost extinct, and the two-horned, all of which are in need of protection if they are going to survive.

Medieval artists depicted the Indian rhinoceros with bolt-heads projecting from its "armor." The skin is thick and tough, but certainly not bullet-proof.

Legal hunting and poaching have reduced the African black rhinoceros to a fraction of its former range and it is now found in greatest numbers in Kenya and Tanganyika. Black rhinoceroses (which are not actually black but dark brownish-gray) are not infrequently exhibited, at least in the larger zoological parks, and even under those conditions of captivity they give an impression of being surprisingly agile and fast on their feet. Experience in Africa bears this out. While their sense of smell and hearing are certainly sufficiently keen, their sight is reputed to be weak, which may account for the unpredictable charges they are said to make. Charges have been timed—probably by people trying to escape from them in cars—at 32–35 miles an hour at the gallop, and 27.2 at the trot. Having on many a spring morning watched our African blacks galloping around their out-of-doors inclosure, tails straight up and the dust flying, I can well credit such speeds.

The giant of the rhinoceroses is the white, or square-lipped, of east Africa. Maximum measurements are presumably based on dead specimens and they are gigantic indeed— shoulder heights up to 6 feet, 9 inches and weights of 3 to 4 tons. The animal's weapons are in proportion: like the black, the white rhinoceros carries 2 horns, the front one usually the longer, and the record one was 62½ inches long.

With all this—great size, great speed, formidable horns— the white rhinoceros is a comparatively mild giant, much more social than the black and, it seems, less given to the violent charges that characterize its relative. We had an excellent demonstration of its placidity when the first two specimens arrived at the New York Zoological Park from the Umfolosi Game Preserve in Zululand in 1962. A crane picked up the heavy crates in which the animals had been transported by sea and placed them carefully in the doorway of the elephant house. Keepers working from the top of the crates knocked out nails and loosened bolts so that the inward end of the crates could be lifted and removed. Much banging and hammering was inevitable and when the crate ends were lifted we would not have been surprised if the animals, frightened and confused, had charged out at full speed and crashed against the steel bars at the front of the stall. Nothing could have been more anticlimactic. Once the end of the crate was removed, the animals stood mildly looking around at the strange surroundings, and then they stepped out—almost daintily—and proceeded to make a leisurely exploration of their new home. In due course they came to the pile of alfalfa in one corner, and both the male and female thereupon ignored zoo staff, keepers, press photographers, and reporters and began munching the hay.

Temperamentally the white rhinoceros—so called because of misapplication of its common name in Afrikaans, *witrenoster*—may be quite different from the irascible black, but it is a cardinal rule that no keeper should ever enter the inclosure of an adult rhinoceros of any species or of either sex. That any black rhinoceros, however quiet it may appear to be, is likely to charge at any time is well understood. In this species the horns are the usual offensive weapons, but while the Indian may use its horn on occasion, it has a real predilection for biting. I once saw a supposedly gentle female Indian rhinoceros savage a steel cage bar with her teeth just after missing the rapidly departing rear of a too-trusting keeper.

Until comparatively recently, births of rhinoceroses in captivity were rare indeed. For one thing, the animals were so costly that few zoological gardens were able to own pairs. For another, even when male and female of the same species were kept, their violent battles were so alarming that they were usually separated to save them from serious injury. It appears that the numerous births in recent years have been due largely to the determination of those in charge to let them fight it out, sometimes with horns carefully blunted. Actually, these brawls between the sexes are often less serious than they appear to be, but the greatest obstacle to successful breeding in captivity continues to be the difficulty of persuading potential parents to tolerate each other long enough for the purposes of procreation. Once a baby arrives, female rhinoceroses have generally proved to be excellent mothers.

THE PIGS

"Pigs is pigs," as everybody knows, and even wild pigs might seem to be comparatively uninteresting as zoological exhibits. This is hardly the case, however, and it is unfortunate that because of import restrictions and the high cost of transportation, wild pigs of any sort are actually rarities in the zoological parks of this country.

From time to time we have exhibited the European wild boar, as well as those from North Africa and Japan. Our Europeans were never as large as the maximum of 350 pounds, but they were impressive animals nevertheless. All our wild boars were kept out of doors, in large runs floored with concrete or asphalt and stone, leaving a small area filled with earth or sand for a wallow—for in their liking of a wallow, wild pigs are little different from the domestic varieties. There was little difficulty about feeding them: available greens, potatoes, carrots, apples, meal, bread, chopped horsemeat, and some alfalfa hay. This regimen was by no means adhered to with regularity, for little that is edible comes amiss to a wild boar.

The rooting ability of the wild boar was demonstrated in spectacular fashion by two North African animals. Starting from some small crevice in the asphalt floor of their yard, they slowly but inexorably demolished the paving, which was three to four inches thick, and reduced it and the crushed stone beneath it to rubble. The yard was unsightly in the extreme but we knew that any new paving would sooner or later be ripped up, and so we resolved to make a virtue of it. A descriptive label was placed at the front of the yard, explaining that pigs like to root, that these had found a tiny crack, and the rubble was the result. It added a great deal of interest to the exhibit for our visitors!

Young wild boars are marked with longitudinal stripes of light and dark gray, and since many of our breeds of domestic swine have been derived from this wild stock, we have been interested to see that domestic piglets obtained from large pig farms for our children's zoo have occasionally been stripe-marked, although recent introduction of the blood of the wild boar seems unlikely.

Because of its grotesque appearance, the wart hog of open or scrub country from Ethiopia west to Senegal and down to South Africa is among the most popular of zoo animals. Grotesque is a mild term for it; I have seen it referred to as a "four-legged nightmare." Grayish in color, it is almost devoid of hair, save for a long, thin mane on neck and back and a tuft at the tip of the tail. Two pairs of protuberances or "warts" on the side of the face give the animal its name. The upper canines curl up, out, and then inward, in a great sweep, and sometimes in males reach a length of more than 2 feet. The lower tusks, seldom exceeding 6 inches, are sharpened by rubbing against a small section of the upper ones, and thus are formidable weapons.

Like other pigs, the wart hog feeds chiefly on roots, grass,

"Grotesque" and "a four-legged nightmare" are terms that have been applied to the African wart hog.

and other vegetable material, as well as on carrion. In feeding, it usually kneels on its forelegs, which have thick pads on the wrists or "knees." It dens in holes in the earth, either digging them itself or taking over from aardvarks or other burrowing animals; a pair kept out of doors here in summer excavated a snug burrow for themselves in a bank near a pool. In entering its den, the animal usually goes backward in order to defend the entrance against possible intruders—and the knife-edged lower tusks are, indeed, excellent defenses. Alarmed when afield, the wart hog makes for its burrow at a smart trot, its tail stiffly erect, and the tail-tuft waving like a flag.

The wart hog accommodates well to captivity and soon loses any tendency to timidity. Many individuals become gentle and a long series of males—or even odd females— kept here and traditionally known as Clarence have endeared themselves to keepers and public.

The wart hog has been bred in captivity perhaps more

often than have most other wild pigs, although successes certainly have not been frequent. To our regret, the homely piglets have never been born in our collection; the male of the only potential breeding pair kept here invariably was seized with convulsions when he even approached the female, so that we had to keep the animals apart.

THE HIPPOPOTAMUSES

Africa is the sole home of the hippopotamus and the pygmy hippopotamus. As zoo animals both have had interesting histories. The Roman Emperor Augustus seems to have kept a hippopotamus in his menagerie as early as 29 B.C., the Zoological Gardens of London received a hippopotamus named Obasch in 1850, and the Jardin des Plantes in Paris got one in 1853—both it and the London specimen being gifts from the viceroy of Egypt. The pygmy hippopotamus, on the other hand, while described scientifically in 1844, was not, with one exception, brought out alive until 1912.

If one wants to accumulate first-hand experience with an animal, there is something to be said for a regular turnover and longevities of medium duration. Until fairly recently our experience with the hippopotamus in the New York Zoological Park was confined to a single specimen, the renowned Peter the Great. Pete was born in the Central Park Menagerie in New York City on July 13, 1903, and was transferred to us on July 14, 1906. He lived with us until his death on February 1, 1953, a span of 49 years, 6 months, 19 days, the greatest then recorded for a captive hippopotamus.

While captive hippopotamuses may sometimes become dangerously savage, as was the case with London's Obasch, Pete remained completely gentle during his long stay in the Bronx Zoo. Toward the end arthritis hampered his movements, but more serious was a condition presumably caused by the wearing down of his molars, combined with reduction in tone of his cheek muscles, so that most of his food intake was lost through the corners of his mouth. We remedied this handicap by grinding his rations and mixing them in a watery gruel that he readily swallowed. This sustained him well until arthritis virtually immobilized him and his painless destruction became necessary. His weight at death was 3,102 pounds.

Pete taught us much, but of course there is a limit to what can be learned from one animal and our education advanced rapidly after Pete's death when we began to maintain pairs of hippopotamuses.

Our first baby hippopotamus was found dead and floating in the hippopotamus pool in the elephant house on December 23, 1958. Four days later the female appeared to be receptive and copulation was observed. This took place in the water, although this is apparently not invariable. Acceptance of the male 4 days after the loss of a calf immediately after birth has been noticed in at least one other zoo.

Further sexual activity was observed in March and April and then, as the months passed without recurrence of estrus, we became assured that another birth was in prospect. Since only a single stall and pool were available, a temporary partition of heavy aluminum posts and bars was erected across the stall, so that we might conform with the established custom of removing the male when a birth is impending. By pushing with muzzle and shoulders, however, and finally by rearing on his hind legs and applying his weight, the male soon so weakened this structure that it had to be removed. The male then promptly rejoined his mate in the pool. Her restlessness increased, and on the afternoon of December 15, 1959, she drove the male up the steps leading out of the pool, and into the stall, whereupon we closed the door at the head of the steps to bar his return to the pool.

On the morning of December 16 it appeared that birth was imminent and the pool was quickly drained, sketchily hosed out, and the valves set for refilling. This had progressed to a depth of about 18 inches and, at 10:05 A.M., while the female was in a standing position, a watching keeper saw the calf unceremoniously dropped into the water with none of the explosive force sometimes reported. The calf swam at once. At

10:35 A.M. it scrambled onto one of the steps, just awash, where it was barely able to stand and was quickly pushed back into the water by a nudge of the mother's muzzle. The calf continued to swim about the pool, apparently excited, but at 12:30 P.M., when the water had reached its maximum depth of 3 feet, the female was seen to be lying on her side and the calf was nursing quietly under water. This situation continued for the next two days, the calf nursing regularly, rising to the surface to breathe at intervals of up to 40 seconds.

On December 18, when the calf was 2 days old, the female was admitted to the stall occupied by the male, the calf following somewhat shakily up the steps. The female was strongly antagonistic toward the male, roaring loudly and finally attacking him by striking upward to bring the heavy lower tusks into play, thus inflicting several superficial cuts in his skin. He defended himself as best he could with open jaws, but made no effort to strike back and appeared only to seek escape. The mother and calf finally returned to the pool, and when introduction was again tried several days later, she seemed somewhat less violent, although still strongly defensive. This situation continued daily until January 20, the animals being separated at night. The female guarded the entrance to the pool vigilantly, and on only one occasion was the male able to enter. The female, with calf trailing, then rushed after him and launched an attack that confined him to a single corner. It was several hours before he could be extricated, uninjured, and presumably greatly benefited by his soaking in the pool.

On the evening of January 19, after the separation gate had been closed for the night, the animals called to each other in low, muffled tones. When the gate was opened the following morning, the female left the pool and approached the male on the floor of the stall. As soon as he rose, she returned to the pool and he followed her. Copulation occurred almost immediately and was repeated at least once during the day. The gate was left open and pairing was again observed, for the last time, on the morning of January 21. Following this interlude, the animals continued to associate on a peaceful

A Nile hippopotamus family.

basis, although the female tried, with gradually decreasing effort, to keep herself between the calf and its father.

The difficulties experienced with our first successful hippopotamus birth might suggest that in spite of the great number of calves that have been reared in captivity, an approved method of handling has not been devised. This is far from true, and most of our troubles stemmed from the inappropriate nature of the only quarters we had available. These quarters were quite satisfactory for the almost 50 years that Peter the Great was our only hippopotamus, but they were not designed for family living.

In general, it is customary to provide adjoining compartments, each with its own pool, so that the male and the female can be kept separately except for mating periods or at least can be separated when a birth is impending. While it is now evident that the depth of the water in which the calf is born is not of great consequence, the general practice in this country is to lower the water in the maternity tank to 12–18 inches, raising it as the calf grows. In Europe it seems to be more usual to allow the calf to be born in deeper water—up to 6 feet.

Too often, when the parents have remained together during a birth, the calf has been lost, a misfortune usually attributed to the father. He may actually have been at fault in some instances, but injury to the calf is more likely to occur accidentally during the turmoil caused by the efforts of the female to expel the male from the pool. Some females are so defensive of their calves that their quarters cannot be approached even by their keepers without the greatest caution.

In our first experience here the male showed no inclination to injure the calf purposely. On one occasion, when mother and calf were on the stall platform and the male had retreated to a corner, the calf escaped the mother's attention and approached the father. It actually managed to touch the great muzzle with its own, which caused the male to back away. This display of caution was quickly justified, for the female became aware of what was going on, rushed between her mate and the calf, and forced the male into another corner.

The hippopotamus is an example of an animal that now occupies a much more restricted range than in earlier times, but is thriving. Once found, in several races, in most of Africa's great lakes and rivers, it has been extirpated from the northern and southern extremes of its range, so that it is now practically confined to the equatorial region, with the greatest concentration in the east. Here it has thrived so well under protection that conservationists are faced with an apparent necessity of reducing its numbers to prevent dangerous overgrazing.

The great, soft body, comparatively large head, and short legs would seem to adapt the hippopotamus to life in the water better than on land, and indeed they do. Both eyes and nostrils, the latter provided with valves, are raised above the general level of the face so that a hippopotamus may float submerged with only these organs and the ears protruding above the surface. It is noticeable to everyone who has seen a zoo hippopotamus close at hand that the skin has glands that secrete a thick, oily exudate, reddish in color. This is produced profusely during periods of excitement and has given rise to the belief that the hippopotamus actually "sweats blood"—a belief that P. T. Barnum, the showman, is said to have encouraged by billing a specimen of the hippopotamus as "the blood-sweating Behemoth of Holy Writ"!

The hippopotamus can stay under water for considerable periods without coming to the surface to breathe, but probably not as long as has sometimes been stated. Numerous careful observations were made in Africa and the normal maximum turned out to be about 4 minutes.

Since the elephant, the rhinoceros, and the hippopotamus are the three largest living land mammals, there is a good deal of interest in hippopotamus weights, and, as usual, a good deal of varying figures and estimates. The figures usually quoted are 2½ to 4 tons, the latter having been given for a male in the Zoological Gardens of London. Figures for hippopotamuses killed and weighed in the field in Kenya run from 5,267 pounds to 5,872 pounds, but most other reports, even from zoo animals, were in the 3,000–4,000 class.

In the wild, aquatic plants form a large part of hippopotamus diet, but the animal habitually makes inland excursions, away from the water, for considerable distances and usually at night, to graze on grass and browse on low-growing foliage. Adult zoo animals do well on 80–100 pounds of alfalfa hay, about 12 pounds of grain mixture in pellet form fortified with vitamin and mineral supplements, and approximately 8 quarts of cut potatoes, apples, carrots, and cabbage, with a loaf or two of stale bread.

The pygmy hippopotamus seems never to have had the extensive range of its big relative, and it is known only from a restricted area in West Africa centering in the hinterland of Liberia, with perhaps extensions into Sierra Leone on the north and the Ivory Coast on the south. As compared with its relative, the appellation pygmy is amply justified, yet it is a more than substantial beast in its own right, for it attains a weight of more than 500 pounds. In appearance the pygmy hippopotamus is strongly suggestive of a young hippopotamus, for in both the head is comparatively smaller than in the adult of the larger species and eyes and nostrils are not noticeably raised above the level of the face. Zoo visitors who so often tell their children that a pygmy hippopotamus is a baby hippopotamus may therefore have some right to their mistake.

There seems to be no record of a pygmy hippopotamus having been seen alive outside Africa until the arrival of a young animal at the Dublin Zoological Gardens sometime in the 1860's according to one account and in 1873 according to another. Whether it lived for "several weeks" or "about five minutes"—again, accounts vary—makes little difference; it was the first. It was not until 1912, when Hans Schomburgk delivered to Carl Hagenbeck, the German animal dealer, five living pygmy hippopotamuses, that the establishment of the species in captivity became possible. Three of Schomburgk's specimens came to the New York Zoological Park in 1912 and were of course the first seen outside Europe.

Schomburgk's description of the pygmy hippopotamus as a "dear sensible little beast" will hardly be supported by most of the zoo men who have dealt with the animal in captivity, for most specimens are irascible, some even savage.

Since those early days a dozen pygmy hippopotamuses have been born in the New York Zoological Park and birth procedure has been seen to vary considerably from that of the big hippopotamus. Births usually occur on land, instead of in the water, although in a 1960 birth the mother dropped the calf into the 4 inches of water in her pool. The calf walked at once, in wobbly fashion, its head and back above water. When the mother was lying down in the water, her very small nipples were under the surface and the baby was obviously unwilling to put its head under water. When we drained the pool, the calf seemed much more content. We finally adopted a plan of providing a shallow bath for an hour only, twice daily, to keep the skin of the animals moist.

Today the pygmy hippopotamus is at a rather low ebb in this country, although it is doing better in some European zoos. Longevity is not bad, although hardly up to the marks set by the large hippopotamus; 39 years, 7 months, 8 days was attained by one of our original trio.

THE CAMELS

The camels—familiar zoo animals that they are—cannot be called wild animals; the Bactrian or two-humped and the dromedary or one-humped are both now known certainly as domestic beasts of burden only, the Bactrian being used largely in central Asia from Afghanistan to China, while the dromedary is a common means of transportation in southern Asia and northern Africa. Untamed herds of both species are sometimes found and are usually considered as merely feral, or reverted to the wild state, living in much the same way as the wild horses of the plains of western United States. There is some evidence, however, that really wild Bactrians may exist in the Gobi Desert area of Mongolia and northern China.

The humps are probably the most characteristic feature of both camels. The backbone does not deviate from its normal course to support these structures, that are merely masses of fleshy tissue, large and full when the animal is healthy and well fed, presumably providing a store that may be drawn upon in times of need. When this occurs, the humps may decrease greatly in size or, especially in aged animals, fall to one side.

It is a firm article of belief in many people that the small cells lining the first two compartments of the camel's stomach are used for water storage, accounting for the well-known ability of the animals to go for some time without drinking. Unfortunately for the belief, the cells apparently have a digestive rather than a storage function. While camels have certainly been known to go without water for as long as 10 to 12 days, such deprivation is not conducive to wellbeing. Experts say that working animals should be watered every third day at least, and daily if possible. In the zoological garden, of course, camels usually have constant access to water.

The points of the body and limbs on which the animal's weight bears in the resting position are protected by bare, leathery pads or callosities—one on the chest, and one on each elbow, wrist or "knee," and true knee. These pads are so appropriately placed that they may appear to have been

Bactrian camels are used as beasts of burden in central Asia and are well built for heavy work on difficult terrain.

caused by abrasion during the animal's lifetime, but actually they are present at birth.

Inoffensive though the camels may seem to be, they nevertheless can be very disagreeable, even dangerous, on occasion. The well-known habit of spitting, which consists of the forcible ejection of saliva and even regurgitated stomach contents, is unpleasant enough, but kicking and biting may be really serious. Camels are not noted for their good nature, as their moans and groans under any sort of pressure or urging testify. At such times they are only too willing to exercise any or all of their offensive abilities.

Both Bactrians and dromedaries are likely to be seen in a zoological park, sometimes purely for exhibition and sometimes used as riding animals to carry children. The Bactrian is stoutly built, with comparatively short legs that are well suited for heavy work on difficult terrain. In winter its hair is

heavy and dense, but much of it is shed in the spring so that
the skin is almost bare, with longer wisps on the humps, neck,
elbows, and tail. One of our males measured 7 feet from the
ground to the top of the forward hump, and a male that had
lived with us for 11 years weighed 1,175 pounds at death.

While in general the dromedary is more lightly built than
the Bactrian and with longer legs, it has been bred through
the centuries with such care that definite breeds have been
developed. These fall into two groups: the light but swift
riding camel and the heavy, plodding "baggager." The weight
of a common working dromedary in the field is 1,000 to 1,150
pounds, but in the zoological garden where there is better
food and less work, greater weights are to be expected—
1,990 pounds for an ordinary female in daily use on our riding
track, for example. Some years ago we acquired two tremendous baggagers and the female eventually attained the surprising weight of 2,255 pounds.

The esthetic qualities of camels are never very high, but
they are seen at their best in midsummer and at their worst in
early spring when the winter coat is being shed and even the
healthiest and best-fed camel presents an extremely moth-
eaten appearance.

PÈRE DAVID'S DEER

It is seldom that the term "the rarest" can be applied to an
animal, at least with confidence that it is, and will remain,
true; today's rarities have a way of becoming tomorrow's
commonplaces when determined efforts are made to get
them. There is no doubt about the term being appropriate to
Père David's deer, however. It is the rarest deer in the world
because it is extinct in the wild state and the few hundred
that keep the species alive are all in parks and zoological
gardens so that their numbers and location are known.

Père Armand David, a French missionary and naturalist,
discovered the deer that bears his name in the spring of 1865.

He just "happened," it seems, to climb the wall that surrounded the Imperial Hunting Park in Peking, and in the distance he saw a herd of about a hundred deer of a kind he had never seen before.

Père David was in constant communication with Professor Milne Edwards, at that time director of the Museum of Natural History in Paris, and in due course he wrote to the Professor:

What characterizes the animals I saw was the length of the tail which was proportionately as long as a donkey's—a characteristic which fits none of the Cervidae I know.

It is also smaller than a moose. All efforts I have made so far to obtain a carcass have been unfruitful. It is even impossible to get a part, and the French Legation does not feel able to succeed in obtaining this curious animal from the Government. Fortunately I know some Tartar soldiers who guard the Park, and I am certain that for a more or less high sum of money, I can get some skins to send you before the winter. The Chinese give this animal the name of Mi-lou, and more often that of Ssu-pu-hsiang, which means "The four characters which do not fit together," because they find that this Reindeer belongs to the stag by his antlers, to the cow by its feet, the camel by its neck, and the mule by its tail.

By bribery and later by diplomacy, Père David not only got skins and skeletons of the deer to send to the museum, but prepared the way for living animals to be sent to Europe. A young pair went to the Zoological Gardens of London in 1869 and there were other arrivals in the next few years in London, Paris, Berlin, and perhaps elsewhere on the continent.

They got out just in time, for during the Boxer uprising in 1900 the remainder of the herd in the Imperial Park were slaughtered by European soldiers.

Fortunately, the eleventh Duke of Bedford, foreseeing the coming tragedy, gathered together the few examples living in Europe and established them at Woburn Abbey. Here, at liberty in the great inclosed park of approximately 4,000 acres, the Père David's deer prospered and the herd increased to 300 head. With as much foresight as his father, the twelfth duke some years ago began establishing small herds in other parks and zoological gardens, for he realized that wars and disease might take a severe toll if all were in one place.

Père David's deer has been called "the rarest deer in the world" because it is extinct in the wild in its native China and is known only from captive herds.

The first examples of Père David's deer to be seen in this country were four young animals picked up as fawns at Woburn Abbey in the spring of 1946 and hand-reared at Whipsnade, through the kindness of the Zoological Society of London. These animals, two males and two females, arrived in New York at the end of the year and were liberated in an inclosure of about 1½ acres. For a few minutes after they were turned out of their shipping crates, it seemed that we were unlikely to exhibit them, for they scattered like quail, ran blindly into fences, and dashed away at every approach of their keepers. Within a week, however, they had settled down, and as they matured in subsequent years there has been a succession of births.

The twelfth Duke of Bedford's policy of judicious distribution has been so broadened that there are now Père David's deer in many European zoos as well as in this country, South Africa, Australia, and even China; four animals were sent to the Peking Zoological Garden in 1956 and began to breed. Thus "the rarest deer in the world" is again established in its homeland—although, of course, as a captive and not as a wild species. It has, apparently, been extinct in the wild for many centuries.

THE GIRAFFES AND THE OKAPI

Africa has produced so many remarkable animals that it would be invidious to choose two as preeminent. And yet— the giraffe is an animal that, once seen, is not likely to be forgotten, and its relative, the okapi, will always compel our interest because of the romantic story of its late discovery.

The giraffes and the okapi, unlike as they appear to be, are nevertheless members of the same zoological family, and both are confined to Africa. Their elongated necks, more exaggerated in the giraffes than in the okapi, contain only the normal mammalian number of seven vertebrae, but a more fundamental distinguishing characteristic is found in the short, unbranched, permanent, skin-covered horns borne by both sexes in the giraffe and by males only in the okapi.

All of the several kinds of giraffe differ mostly in their markings that consist of blotches, ranging from pale yellowish brown to nearly black, surrounded by a network of white or yellowish tone. They range over most of Africa south of the Sahara, except the great rain forests of the west. The preferred habitat is dry, open wooded areas or tree-dotted plains, where the great height of the animals enables them to feed on acacia leaves, the principal item of their diet. Giraffes usually live in groups of females and young, dominated by a single adult male, the herd numbering perhaps a dozen animals. In some more remote regions giraffes still persist in fair

Giraffes are much better constructed for browsing on high foliage than for feeding from the ground. This is the animal's usual stance when drinking also.

numbers; in more settled areas they have disappeared entirely or are found only in the great national parks or preserves.

The height giraffes may reach is naturally a matter of great interest. It is known to be greater in males than in females, but accurate figures are difficult to obtain. In order to satisfy the usually incredulous curiosity of our visitors and to gauge our own judgment of height, our indoor giraffe stalls carry at the front a plainly marked perpendicular scale of inches and feet, against which reasonably accurate sighting measurements of standing animals can be made. The greatest height recorded here by this means is 15 feet, 6 inches for a male Nubian giraffe. Giraffes in nature, however, have been recorded from 16 feet to 19 feet, 3 inches for males and an average of 15 feet, 6 inches to 16 feet for females.

Estimates of weight are given as one to two tons for males, but, as with most large mammals, actual weights are seldom available. One female Masai giraffe weighed in the field made 1,760 pounds and an adult male Nubian giraffe in our collection weighed 1,650 pounds at death.

The extremely long neck and high shoulders of the giraffe make it impossible for the animal to drink or to feed from the ground without spreading its legs sideways or fore and aft. This is accomplished with some difficulty, although apparently with no concern. Giraffes in nature appear to go for considerable periods without drinking, and in fact are said to avoid large bodies of water. Rivers seem to provide impassable barriers and it has been said often that they are poor waders and unable to swim. Apropos of this point, in 1960 a giraffe that had just been unloaded from a ship at a New York dock escaped from its crate, ran to the end of the pier, and fell into the water. Observers reported that it sank from sight almost at once without making effective efforts to swim.

While the giraffe in nature has few enemies, its chief protection, when needed, lies in flight. At close quarters the giraffe, awkward though it may seem, is able to defend itself with powerful weapons. It can kick or strike either forward or backward with its heavy, sharp-edged hoofs, while the great head can be swung with lethal force, the effort appearing to be directed toward the delivery of a blow with the side of the head, rather than with the horns. I once saw a bull giraffe strike a female on the lower back, in this manner, with such force as to drive her to the ground.

The giraffe, though often considered mute, is actually capable, apparently under stress, of making small vocal sounds. Some instances have been quoted of a giraffe that blared like a calf when captured and of a cow giraffe that "occasionally gives vent to a low call note when its young strays too far away." A Masai calf that was born here repeatedly uttered a soft, rumbling bleat when separated temporarily from its mother. The latter invariably became agitated at such times but was never heard to make a vocal response. A definite instance of the uttering of a vocal sound by a female reticulated giraffe in our collection has been recorded. This animal, suffering from arthritis, was unable to rise from the usual lying position, with legs folded beneath her. When an attempt was made to lift her, she emitted a low mooing sound, distinctly heard by all present, including myself.

Whether or not the giraffe lies down to sleep has long been a controversial subject. Many observers have agreed with the view that their long legs are such that normally they do not lie down to rest but remain standing all the time, even when sleeping. Observations in nature are not easily made, since a sleeping giraffe is difficult to approach. Even in the zoological garden the situation is much the same, for although accustomed to captivity, it always retains its natural timidity. Caution and persistence, however, have enabled observers to determine that captive giraffes, at least, do lie down to sleep, and photographs have been taken of what appears to be a young giraffe with its head folded over its back, presumably in deeper sleep than other giraffes that sleep while lying down but with head erect.

This "deep sleep" position is said to be maintained for an average of only 6½ minutes perhaps five times during the night.

The mere act of lying down with head erect during daytime can often be observed in zoological gardens where the animals are undisturbed. I have frequently seen our own giraffes in this position in their outdoor yards. Whether or not they were sleeping I cannot say, but one of our keepers reports seeing an adult male Masai lying with his neck folded over his back. All three of our present adults have been seen lying down in their stalls at night with heads erect, but usually an approach results only in the sounds made by the rising animals, long since alerted.

Although the giraffe was known to the ancient Egyptians as a captive animal as early as the Eighteenth Dynasty and to the Romans at least during the reign of Julius Caesar, it was not until the nineteenth century that it became fully established in European zoological gardens. In 1826 a single giraffe arrived in France, in 1827 another reached England, and in 1828 a third was received at Schonbrunn in Austria. It seems probable that a giraffe received in 1873 at the Central Park Menagerie in New York was the first to reach America. We have had them constantly at the New York Zoological Park since 1903.

Usually, animals that have become especially adapted to particular feeding habits are diffcult to care for in captivity, but giraffes are not among them. They readily accept substitutes for their natural diet. The average daily ration for an adult animal is approximately 20 pounds of alfalfa hay, 10 pounds of feeding pellets and 5 pounds of rolled oats mixed with cut apples, carrots, potatoes, cabbage, and bread. Bananas are reserved as a special treat and are fed by hand. Hay we provide in racks hung with the bottom about 7 feet from the floor, and in summer fresh leafy branches are hung on the fences, a treat that is much appreciated, for the leaves are stripped from any low-growing tree in the inclosures as soon as they appear.

Fortunately for zoological garden budgets, giraffes breed well in captivity, although the young are not always reared. If the calf is normal, the usual course of things after birth is for it to achieve a standing position, after several attempts. It then wobbles uncertainly to its mother, fumbles about for the udder, and, sometimes assisted by nudges of the dam's head or forefeet, eventually finds it. Once this has been accomplished, there is usually no further difficulty, as the giraffe cow is generally an excellent mother.

Modern drugs and methods of medication are sometimes all-important, as they were in 1962 when a calf was born to a Masai female in the Bronx Zoo. The calf, a male, was born after an observed gestation period of 457 days, stood 67 inches, and was obviously sturdy for he was soon on his feet and searching for his mother's nipples. She was reluctant to permit him to nurse, except intermittently, and by the next day her udder was swollen and apparently tender, for she avoided the calf's efforts to approach. Our veterinarian thereupon injected a tranquilizer by means of a Capchur pistol, a device that projects a drug-loaded hypodermic syringe, and a diuretic was given in the drinking water to reduce the swelling of the udder. This medication was fully effective and on the following day the calf was nursing again and continued to do so normally.

Births of giraffes are usually of single young, although twins have been reported in a few instances. Height of the calves at birth shows some small variation, from 66 to a little more than 72 inches, and weight also varies; we have recorded it from 87 to 107 pounds. Our best longevity is about a month under 22 years, but the Zoological Gardens of Antwerp have kept giraffes for as long as 28 years.

The furor created by the discovery of the okapi in 1901 left an aura of mystery that still surrounds this beautiful animal. That a creature so large and, once in the open, so conspicuous, could have remained so long unknown to any but the primitive inhabitants of the steaming jungles of the Congo even now seems almost incredible. The explanation, at least in part, certainly lies in the nature of the okapi's home.

As far back as 1918 the New York Zoological Society made a determined effort to bring out an okapi and a newborn calf was picked up for us by the Congo Expedition of the American Museum of Natural History, but the expedition's supply of condensed milk was soon exhausted and the baby died after 10 days. Herbert Lang, who headed that expedition, wrote a graphic description of the okapi country for the Zoological Society's magazine:

Whoever penetrated here was, so to speak, "on the wing," and wings beat doubly fast across these inhospitable regions. The numerous sportsmen who had visited nearly all parts of Africa found no attraction in these forests. Indeed, the many pale, haggard faces that emerged from the western half of equatorial Africa were no incentive to pleasure-seeking people. The immensity of the wilderness is appalling; for over eighteen hundred miles without a break it stretches more than half way across the continent. In spite of tropical luxuriance, it is one of the most dismal spots on the face of the globe, for the torrid sun burns above miles of leafy expanse, and the unflagging heat of about one hundred degrees, day and night, renders the moist atmosphere unbearable. Over the whole area storms of tropical violence thunder and rage almost daily. Here natives have become cannibals, and the graves of thousands of white men are merely a remembrance of where youthful energy and adventures came to a sudden end.

The discoverer of the okapi was Sir Harry Johnston, special commissioner in Uganda. In 1901 he had occasion to go into the (then) Congo Free State to restore to their homes a band of pygmies that had been captured by a filibustering German. Johnston had read Stanley's *In Darkest Africa* and recalled a mention by the explorer that the pygmies sometimes captured in pits an animal they called "atti," so when he was in pygmy territory he made inquiries of Belgian officials he encountered. At the Belgian post of Mbeni he had luck; the Belgians knew of such an animal, and in fact thought there might be a skin lying around. As it turned out, the skin had already been cut up by the native soldiers for belts and bandoliers, but Sir Harry obtained two strips of the hide and sent them to the Zoological Society of London for examination. The strips happened to be from the vividly striped flanks of the animal, and were so zebra-like in their markings that they were at first thought to be from a hitherto undescribed zebra. Eventually Sir Harry was able to send out a skin and two skulls and when these reached London they were recognized as having an affinity with the giraffes rather than the zebras, and so a new genus, *Okapia,* was created with the specific name of *johnstoni* in honor of Sir Harry.

It was only a year after the failure of our attempt to get a young okapi in 1918 that a successful effort was made. An okapi calf was hand-reared by the wife of a district commissioner in the Congo and reached the Zoological Gardens of Antwerp in August, 1919. It lived only 51 days, but subsequently okapis came out of the Congo with some frequency, especially after a station for their capture, study, and supply was established in 1946 at Epulu in the Ituri forest.

Conditions have of course changed in the Congo since Lang's day and the okapi country is no longer quite as forbidding as it was in the second decade of the century. Indeed, well-surfaced roads now give easy access to the dark forests where the okapis live.

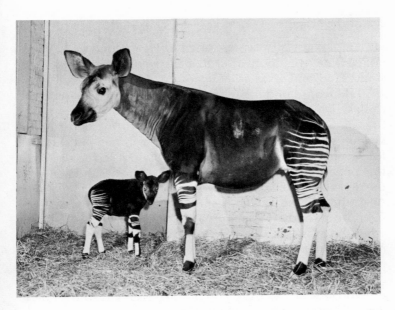

The okapi was discovered in the Ituri Forest of the Congo in 1901 but was not seen in captivity until 1919. Now pairs are in several zoological gardens, and births are not uncommon.

It was not until 1937 that we received an okapi, a lovely male that had been captured in a pit and bottle-reared by Brother Joseph Hutsebaut at the Mission of the Premonstratensian Order at Buta in the northeastern Congo. Since then we have received two more from the Congo, and three have been born here, although only two were reared.

Zoo men are accustomed to most animals births and have learned that the best procedure is to let nature take its course without interference. We were not, however, quite able to be so insouciant and unconcerned in 1959 when the first birth of an okapi in our collection was impending. Since the only statements of the gestation period of the okapi were at wide variance—from about 10 months to 426 days—we could only wait, from the tenth month on. Our expectant mother, a gentle okapi named Muyoni, occupied one-half of a double stall in our antelope house and in preparation for the birth the adjoining stall was darkened and provided with peep-holes, while the floors of both stalls were well padded with straw, sand, and dry, sterile sugarcane residue. By October 15, 1959, Muyoni's udder had increased in size and she became restless, lying down for a few seconds, then moving from stall to stall. Soon after midnight the curator of mammals and a keeper, who had been keeping an all-night vigil in the dark for several evenings, were able to make out in the dim light that one foreleg of a calf had appeared. Muyoni then entered the darkened stall and lay down. When she rose again at 1:45 A.M., the faint beam of a flashlight revealed that the calf had been born. Forty-five minutes later he was on his feet and when he was less than three hours old he had found his mother's nipples. We were able to work out the gestation period exactly—441 days in this case.

Okapis live reasonably long lives in captivity; Antwerp kept one for 15 years, 40 days, and our first okapi lived with us 15 years, 33 days. There are now so many okapis in the zoological gardens of Europe and America that the future of the animal in captivity, whatever may be its fate in the wild, seems assured.

THE BONGO

Sometimes it is difficult to be calm about the beauty of an animal. Surely the bongo must be numbered among the rarest and least well known of the antelopes, as well as among the most beautiful—indeed, the most beautiful, in my estimation.

Colonel E. Percy-Smith caught our first and only bongo—which was also the first to be exhibited alive—in the Aberdare Mountains of Kenya in 1932 and in January of the following year she arrived in New York. Doreen, as Colonel Percy-Smith had named her, had been captured in a snare set along well-used game trails high in the mountains.

"Doreen," the first bongo to be exhibited alive, came to the New York Zoological Park from Kenya in 1933.

"I was aroused at crack of dawn by a great hubbub in camp," he wrote. "The scout had hurried in to report that a bongo had been snared. In the thrill of success it seemed no time before I had covered the mile that intervened, and there, sure enough, to my intense relief and delight, I saw a most beautiful half-grown female bongo securely held round the neck. Fortunately, she did not show so much fear as I had anticipated; I was able to approach her without undue difficulty, and, with the help of a dozen natives, to rope and put her bodily into the back of my big 7-seater tourer."

Colonel Percy-Smith found that Doreen was so docile he did not need to keep her in semidarkness at his camp, and he gave her as companions a cow and calf. She and the calf took turns nursing. Doreen's captor wrote:

It is a perpetual interest to be able to study this most attractive creature at close quarters, and to admire the gorgeous chestnut hue; the unexpected thick ridge of hair running along the spine—nature's protection and warning against the overhanging boughs of her natural haunts; and the white transverse stripings of the coat, all the more curious because, for some mysterious reason, there is one less on one side than on the other. But there, her fascinations have no end—at least for me.

They had no end for us either, for Doreen was the ideal antelope exhibit—beautifully colored and marked, with large, limpid eyes, always gentle and trusting. She lived here peacefully until her death after 18 years, 2 months, and 19 days in the collection.

Gentleness seems to be a characteristic of the bongos, for many years later when another specimen was captured for us in the Congo—this time an adult male—it became tame enough after a week in a stockade to come up to anyone for proffered food. This animal was lost unfortunately through mishandling by native carriers. Sometime, it is to be hoped, bongos will be sufficiently plentiful in zoological gardens so that a small continuing supply can be expected from breeding, but that day has not yet arrived; less than a dozen have so far been successfully captured and brought out. And we still, after all these years, know little about its life in the wild. Apparently it ranges in forested areas from Sierra Leone to western Kenya, staying closely within deep forest where it is difficult to observe, and feeding chiefly on foliage. The Cleveland Zoological Park now has a young pair, the only ones in this country, and would certainly have no trouble disposing of the calves if any are born!

THE BISONS

On December 8, 1965, with appropriate ceremonies, the New York Zoological Society unveiled a tablet on the south wall of its lion house commemorating the sixtieth anniversary of the founding of the American Bison Society, the organization that took a leading part in assuring the continued existence of the largest—and at one time perhaps the most common—of North American mammals. It was in the lion house that a group of conservationists met under the presidency of Dr. William T. Hornaday, at that time director of the zoological park, to arouse public interest, to acquire the few available bison, and, in short, to save the animal from extinction, by any means possible.

The young zoological park had already begun collecting bison and had acquired four bulls and three cows from C. J. Jones, better known as "Buffalo" Jones, in October, 1899, just before the zoological park was opened to the public in November. The bison was at a low ebb at the time, for the great herds of the western plains had long since been shot off, and it was obvious that the few remaining animals in private hands should be grouped and established in favorable areas under protection. By late 1903 we had some forty animals on hand.

In 1905 the New York Zoological Society offered to the government of the United States a nucleus herd of bison for installation in the Wichita Forest Reserve in Oklahoma, provided that a suitable area would be fenced. The offer was accepted, a range of about 6,200 acres was inclosed, and fifteen animals were shipped out by rail. With the exception of four bulls from the Fort Niobrara Wildlife Refuge in Nebraska, no additions to the original stock were made and the herd thrived so mightily that the range had to be expanded to 59,000 acres. Fifty years after the establishment of that herd, it numbered 973 animals, more than 3,600 calves had been born, and more than 2,000 bison had been slaughtered or released to zoological gardens.

The Zoological Society helped also in the establishment of

the bison herd in the Wind Cave National Park in South Dakota, sending fourteen bison there in 1913. After that, our bison herd was allowed to wane, for the principal reason for its establishment had been satisfied.

In its heyday, our bison-breeding project was an undertaking of some magnitude, one that could not ordinarily be maintained by an urban zoological garden. The main range embraced an area of at least 10 wire-inclosed acres of open grassland. Adjacent but concealed was a low, unheated shelter, suitably divided into useful compartments. It adjoined a large feeding corral, with many scattered feeding boxes placed on the ground and hay racks under the eaves of the shelter. The herd was drawn into this inclosure after the food had been put out and usually was not released until morning. There was also a strongly reinforced bull pen, generally containing a huge male that had become too overbearing to be allowed to run with the herd. On the higher ground of the main inclosure the bison excavated a number of wallows, used year after year. In dry summer clouds of dust rose from the area as the animals rolled in the fine dry earth. During periods of rain the wallows filled with water, providing mud baths as a welcome respite from biting flies. Except in the severest winter weather, and then by no means routinely, the bison seldom used their shelter house and the sight of twenty or thirty bison standing patiently as snow drifted down and whitened their backs, must have given many visitors to the zoo a faint but unforgettable impression of what the great bison herds were like in the old days of the west.

All the bison that went out from the New York Zoological Park were plains bison and we have never exhibited the subspecies known as the woods bison. This animal was once fairly well distributed in the western United States and Canada, but by 1891 it had been reduced to about 300 animals in the Great Slave Lake area. It was then taken under protection by the Canadian government and the herd increased so that, like the plains animal, it now seems safe for the foreseeable future.

*Every pure-blood wisent, or European bison, is registered
in a Pedigree Book and is given a name indicating
the herd from which it came. All wisents born in the
New York Zoological Park have
names beginning with "Ne."*

Incidentally, there is one perennial question concerning the American bison that plagues the staff of the New York Zoological Park. It has to do with a bison named Black Diamond supposed to have been the model for the buffalo nickel. The National Archives and Records Services in Washington has reported that "a search of records pertaining to the buffalo nickel that are in the National Archives among records of the Bureau of the Mint (Record Group 104) failed to show any mention of 'Black Diamond' or any identification of the buffalo that served as a model for the buffalo on that nickel." Our own records throw no light on the question; no Black Diamond is listed on the file cards that so carefully record data for the period when we were in the bison business on a large scale. A bull named Black Dog was born in the zoological park and was included among the fifteen animals sent to the Wichita Game Preserve in 1907, but there is no mention of its having served as a model for the buffalo nickel. All we know is what we can tell by looking at the reproduction of the coin—the model was a magnificent specimen.

Shoulder height of the plains bison is given as 5 feet, 9 inches, the females being considerably smaller. Weight is usually given as about 2,000 pounds for males and 1,000 pounds for females, but these figures are sometimes exceeded by exceptional animals—a woods bison was accurately weighed at 2,402 pounds, for example.

We have never attempted any hybridization, but bison and domestic cattle interbreed readily, domestic bull on bison cow being more successful than the reciprocal mating. The production of such hybrids, known as cattalos, has proved not to be commercially worth while, largely because of limited fertility.

Long-lived—the best record is just 3 days short of 26 years —and impressive in appearance, with an aura of the old west inevitably clinging to it, the American bison will always be one of the great exhibits of any zoological park.

Subjective judgments of animals are perhaps inevitable, and so it must be reported that there is not universal appreciation of the massive bulk of the American bison. Some

Europeans take a different point of view. I am thinking of a well-known French zoologist who commented, after we received our first European bison, or wisent, in many years: "You perceive the difference, of course: the wisent is a Cadillac; your American bison is a Mack truck!"

I did *not* agree, but it is true that the wisent is a taller animal than the American species, with shorter mane and beard and less pronounced development of hump and shoulders, giving it a more rangy—a more Cadillac!—appearance. Certainly it is not noticeably smaller, for a weight of 2,001 pounds has been recorded, and a shoulder height of 6 feet, 2 inches.

Once found over most of Europe with possible extension into Siberia, the wisent was gradually eradicated from much of its range, increasing development of the land and destruction of the forests bringing about the inevitable result. The typical or Lithuanian race became restricted to the Bialowieza Forest of Poland, where, in 1803, a herd of 300 to 500 came under the protection of the Czar of Russia. Then followed a series of fluctuations and disasters, caused by wars and political disturbances and undoubtedly the over-browsing of the forests. The Bialowieza herd was exterminated by 1921, and it was not until 1929 that efforts were made to reestablish it. Almost concurrently, the few remaining wisents of the somewhat smaller and darker Caucasian race, living in the mountains of southern Russia, met a similar fate; the last of the Caucasian race was killed in 1927.

Fortunately, in better times numerous wisents had been sent from Bialowieza to various zoological gardens and private preserves, so that at the end of World War II at least these scattered animals and the remnants of the rebuilt Bialowieza herd remained alive. Then began the task of locating the animals, establishing their pedigrees and taking whatever steps could be taken to insure the preservation of the species. Today there are records of between 300 and 400 pure-blood wisents, and much of the credit for this apparent success must go to the International Society for the Protection of the European Bison, organized in 1923 and modeled after the older American Bison Society. In spite of a rather uneven career, the Society has managed to maintain its records and to publish its *Pedigree Book,* some seven issues of which have appeared under European editorship.

A serious obstacle to clearing the eligibility of individual animals for inclusion in the *Pedigree Book* has been the frequent crossing with the American bison. Such hybrids are readily produced, of course, and at one period, when the plight of the wisent appeared hopeless, interbreeding appeared the only recourse. But these hybrids have been rigorously excluded from the list of pure-bred animals, and now there are enough of the latter so that there is no longer interest in the production of hybrids.

The first European bison to be seen alive in this country seem to have been a pair received at the New York Zoological Park in 1904, purchased by a friend of the Society after receipt of a postcard offer from the Duke of Pless, whose herd originated with stock obtained from Bialowieza in 1865. The bull lived for just over 8 years, the cow for 13, but no calves were born.

It was not until 1959 that we received more, a pair sent to us as the gift of the Zoological Society of Amsterdam. Every registered wisent has a name, the first two letters of which are keyed to its place of birth. "Ar" is the designation of "Artis," the Zoological Gardens of Amsterdam, and so our pair were named Aristo and Ardetta. Our designation is "Ne," and so when our first calf was born it was named Nerissa. A second calf was called Nerita. We can foresee plenty of naming problems as the calves grow up and attain breeding age, and the herd builds up.

Taken overall, there seems to be good reason for accepting the belief of European zoologists that the future of the European bison is assured, at least as far as that of any species can be assured in captivity. In 1952 two animals were returned to the Bialowieza Reserve in complete liberty and eight years later a group of more than a dozen freely roaming wisent had been built up. But unless the forests of Europe themselves can be restored and protected, the future of the wisent as a wild animal, living in a state of nature, seems dim indeed.

THE ARABIAN ORYX

Remembering the heroic efforts that had to be made to pre-serve the American bison, we can still be thrilled today to realize that equally heroic measures are being applied to other animals in dire straits. The Arabian oryx is one of them, and since 1962 it has been almost constantly in the world's news.

This is the smallest of the oryxes, with a shoulder height of about 35 inches. The body color is white, with patches of blackish brown on the forehead, nose, and cheeks. Horns are short and almost straight, reaching a record length of 27¼ inches.

The Arabian oryx once had a fairly extensive range in the desert areas of the Arabian peninsula and nearby territory, but constant hunting, in recent years by means of motor cars and even airplanes, has reduced the animals to perhaps 200 in a remote and desolate area of southern Saudi Arabia.

In earlier days, when we knew the Arabian oryx as the beatrix antelope, it was a fairly common zoological garden exhibit. It lived and bred well, but unfortunately no one then realized the necessity for establishing it firmly. As a matter of fact, between 1903 and 1906 we acquired a pair and in due course four young were born. Eventually they all died, but we were not particularly concerned, for the species seemed to be a standard zoological item, obtainable readily enough through the usual channels of dealers and animal collectors.

As the years went on, however, Arabian oryxes did not come into the animal market, and disturbing reports began to arrive of the decimation of the wild stock. The situation grew so bad that in 1962 a project for the capture of speci-mens for transfer to a protected area was organized by the Fauna Preservation Society, and as a result two males and one female were captured and safely transported to temporary holding quarters in Kenya. Later, these three animals, with a female supplied by the Zoological Society of London, were deposited with the Arizona Zoological Society at Tempe, Arizona, where the hot, dry climate was thought to approxi-

The Arabian oryx is so nearly extinct that efforts are being made to breed it in captivity. This calf was born in the New York Zoological Park in the early years of this century, before the species was endangered.

mate conditions in their native habitat. Other animals from private sources in Arabia were added, births followed in due course, and now there is at least a nucleus captivity herd, whatever fate may befall the few remaining animals in the wild.

THE TAKIN AND THE MUSKOX

For many years a woodenly mounted and to tell the truth rather moth-eaten takin gazed at visitors from a glass case in the Heads and Horns Museum in the New York Zoological Park. Natural history museums of course had other and perhaps better-mounted specimens, but no zoological park, at least in America, could boast even a stuffed takin. Ours was a distinction we were not very proud of, however, and the glassy-eyed, calflike figure served mostly to remind us that some day, somehow, we must have a real, live takin.

Unfortunately, the takin, whose closest relative is the muskox of the Arctic regions, lives in the mountainous areas of southeastern Asia, often above tree line, from Bhutan northeastward to Shensi in central China, and this is not an area in which animal dealers abound or collecting expeditions often operate. Only two living takins had ever come out of Asia, both to the Zoological Gardens of London in 1909 and 1923.

In 1958 it happened that the New York Zoological Society sponsored a Burma Wildlife Survey being made by Oliver M. B. Milton and Richard D. Estes and, in the course of discussions about their project before their departure, we mentioned that if they happened to run across a takin that was available for sale, we would be interested to hear about it. It was an off-chance, of course, but the two men would certainly be in takin country for months and just might happen to hear of something.

In midwinter of 1958–59 cables and letters began to flow in from Burma. Mr. Milton had mentioned our desires to government officials and to his surprise was told that there *was* a takin in captivity, a very young female that had been picked up by hunters of the village of Sankawng in northern Burma the previous July. As a protected animal, it belonged to the Kachin State government, but it was for sale at $5,000. Mr. Milton gathered from the conversations that if the Bronx Zoo bought the animal, the money would be used to build a schoolhouse and a road connecting two villages.

The price was high, but the takin had a reasonably good longevity—London's two specimens lived approximately 9 and 12 years, respectively—and no live takin had ever been seen in the New World. We accepted by cable and Mr. Milton and Mr. Estes in January flew as close as they could to Sankawng and walked the rest of the way to inspect the animal and take photographs to send to us.

They found that it was a village pet, running at complete liberty with the village cows, goats, and dogs, getting into gardens and making a nuisance of itself by nibbling young plants. Four Nung hunters had caught it in the snow-covered mountains nine days' walk from their village and had taken turns carrying it out on their shoulders—not a heavy burden, apparently, for it was only about two months old at the time. It was so tame and friendly that they never thought of penning it up when they reached the village and merely liberated it with the rest of the livestock—a circumstance that horrified Milton and Estes, for they knew that tigers and leopards roamed the nearby forest and indeed, on their first night in the village, they were awakened by the barking of the dogs when some prowling animal came close and villagers rushed out with torches and loud cries to scare it off. They were assured, however, that the takin was perfectly safe, for at nightfall it always bedded down under one of the houses in the center of the village.

Some weeks passed before air transport could be arranged from the airstrip nearest Sankawng to Rangoon and Mr. Milton returned to the village to supervise the departure. The takin had become strongly attached to one of the village cows and so to encourage it to leave the village and walk the several miles to the airstrip, a procession was formed with the favorite cow leading the way. After the first half mile the cow had had enough and bolted, but the takin docilely followed the village men and kept plodding on. It was the beginning of the hot season and whenever they passed a roadside brook, the villagers stopped to bathe the little animal. On the open road, one of them held an enormous umbrella over it to protect it from the sun.

Rangoon was even hotter and difficulties multiplied, for a takin has a musky odor and no airline wanted to carry it in a passenger plane, the only transport available at the time. For some weeks the takin was kept in an air-conditioned room at the airport and later in a fairly cool corner of the Rangoon Zoo, but except for losing a little weight the takin withstood the heat quite well and was frisky and healthy when she was

Mrs. Oliver Milton and her son Eric, who had played
with the young takin in Burma, greet it on its
arrival in the New York Zoological Park.

finally put aboard a plane for the flight to New York. After
thirty days in quarantine, she was released to the New York
Zoological Park on October 30, 1959, and went on exhibition
that same afternoon.

Through the slats of the shipping crate we could see that
she was a stout-bodied, short-coupled animal with high
shoulders and comparatively short legs, largely blackish and
with the back, except for a dark stripe, being grizzled brown.
The musky odor that had so worried the air line was noticeable but, to most of the Zoo's assembled staff, not particularly
unpleasant. In an airtight airplane we might have thought
otherwise.

Mrs. Milton and her son Eric who, as a boy of four, had

played with the takin in Sankawng during a visit several months before, were at the zoo to welcome their Burmese friend. Mrs. Milton assured us that the little animal would not be obstreperous, but everyone stood well back and ready to flee when the end of the crate was removed.

The takin stepped out, gazed around, sniffed the ground. Mrs. Milton extended a handful of hay and the animal strolled up to her and fed while she scratched its ears. Little Eric patted his friend and asked if he could ride on her back, as he had done in Sankawng. It was just about the most peaceful and uneventful arrival of a great rarity that I can remember.

That afternoon we bestowed a pet name on our first takin, calling her Gracie after Miss Grace Davall, the Assistant Curator of Mammals and Birds, who had singlehandedly carried on the almost interminable negotiations that finally got the takin aboard an airplane from Rangoon to New York. With the winter ahead of us we had no fears about Gracie not being able to withstand New York weather, for she was completely indifferent to cold and snow. By the following summer we had outfitted an air-conditioned stall inside the antelope house, however, and here she spends the summer months except on unusually cool or overcast days. Her weight on arrival was 155 pounds and during her stay in the Zoo she has obviously grown considerably, although we have had no opportunity to weigh her. It would not be easy; long out of calfhood, she is still frolicsome and inclined to alternate between stolidity and skittishness. Male takins are said to reach weights of 600 to 700 pounds, females being smaller. Gracie has a good appetite for the food we give her—8–10 pounds of alfalfa hay, about 3 pounds of feeding pellets, and rolled oats, bananas, apples, carrots, sweet potatoes, cabbage, and fresh foliage in summer, and it will not surprise us if she eventually reaches maximum size for a female.

To link the takin of Burma with the muskox of the Arctic may seem strange, but their zoological relationship is generally accepted. They have another point in common: the muskox is almost, if not quite, as rare in zoological parks as the takin.

As an outstanding North American animal, the New York Zoological Park had made vigorous efforts to obtain it in time for the opening of the zoo in 1899. The experienced animal man, "Buffalo" Jones, managed to rope and capture five yearling calves in the Canadian Barren Grounds for the zoological park, but on his way south Indians stole into his camp at night and slaughtered all of them "to keep the other muskoxen from following them out of the country." We finally got one in 1902, the survivor of four calves captured the year before; the other three had been killed by sledge dogs.

At intervals over the years, a few others have come into captivity here and abroad, but few have survived more than half a dozen years. Still, in May, 1953, the body of a female muskox found floating off Nunivak Island, Alaska, bore an ear-tag indicating she had been born in Greenland in the spring of 1930, so longevity is potentially good. As zoological gardens learn more about diets and disease, and as animal management approaches a science, no doubt the muskox, like so many other once-difficult animals, will become familiar to every zoo visitor.

A NOTEBOOK NEVER ENDS

Squirrels are known to gather and store food far beyond their momentary needs, and I feel a kinship with them, for I have been collecting observations on wild animals for almost six decades. Even now, scarcely a day passes without another reference or another peculiarity of behavior, or diet, or reflection, being added to my long row of index cards. I would like to think that whereas a squirrel's hoarding may be only an innate compulsion, *my* garnering is purposeful and may be described as a grain-by-grain addition to the art and practice of zoo-keeping. But whatever it is, it has been an enormous satisfaction all my life.

And yet I must admit that the deepest satisfactions of a

long life in the zoo are not those set down in notebooks. How can you—or why should you?—make a few penciled notes to record how you feel as you await the return of a collecting expedition? Or the quiet, glowing joy of staring for the first time at an animal rarity you never expected to see alive? You don't enter those pleasures in a notebook to keep from forgetting them. You won't, no fear.

It has been a long time since 1908 when I began working with animals in the New York Zoological Park, and inevitably, looking back, I ask myself what was the mainspring of all those years of work, worry, success, failure, accomplishment. I suppose it was acceptance of the challenge that every captive wild animal presents. It is almost as if the animal said in so many words: "All right, you've taken me out of my natural environment. Now it's up to you to keep me alive and healthy and to give me a good life in your zoo."

That is not an easy challenge to meet and I don't pretend that we invariably met it successfully in earlier days, or that we do so now, although we certainly come a lot closer. I think that's what sustains and motivates a zoo man—the coming ever closer to understanding the nature and needs of the wild animals under his care, devising ways of satisfying them, seeing the animals respond to his perceptions and ingenuity.

Essentially the zoo man follows his profession because he likes animals and wants others to like them too. I know it is customary to justify the maintenance of wild animals in captivity—a zoo, in other words—by saying that from the study of these lower forms we learn about ourselves. No doubt there is a great deal of truth in this, but I do not need it for my own justification of a life as a zoo man. I have had a thousand better justifications over the years as I have watched people enjoying well-kept animals in the zoo.